A Logical Foundation for Potentialist Set Theory

In many ways set theory lies at the heart of modern mathematics, and it does powerful work – both philosophical and mathematical – as a foundation for the subject. However, certain philosophical problems raise serious doubts about our acceptance of the axioms of set theory. In a detailed and original reassessment of these axioms, Sharon Berry uses a Potentialist (as opposed to an Actualist) approach to develop a unified determinate conception of set-theoretic truth that vindicates many of our intuitive expectations regarding set theory. Berry further defends her approach against a number of possible objections, and she shows how a notion of logical possibility that is useful in formulating Potentialist set theory connects in important ways with philosophy of language, metametaphysics and philosophy of science. Her book will appeal to readers with interests in the philosophy of set theory, modal logic, and the role of mathematics in the sciences.

Sharon Berry is Associate Professor of Philosophy at Ashoka University. She has published articles in journals including *Philosophical Studies*, the *British Journal of Philosophy of Science* and *Analysis*.

A Logical Foundation for Potentialist Set Theory

SHARON BERRY
Ashoka University

CAMBRIDGE
UNIVERSITY PRESS

CAMBRIDGE
UNIVERSITY PRESS

Shaftesbury Road, Cambridge CB2 8EA, United Kingdom

One Liberty Plaza, 20th Floor, New York, NY 10006, USA

477 Williamstown Road, Port Melbourne, VIC 3207, Australia

314–321, 3rd Floor, Plot 3, Splendor Forum, Jasola District Centre, New Delhi – 110025, India

103 Penang Road, #05–06/07, Visioncrest Commercial, Singapore 238467

Cambridge University Press is part of Cambridge University Press & Assessment, a department of the University of Cambridge.

We share the University's mission to contribute to society through the pursuit of education, learning and research at the highest international levels of excellence.

www.cambridge.org
Information on this title: www.cambridge.org/9781108994880

DOI: 10.1017/9781108992756

First published 2022
First paperback edition 2024

A catalogue record for this publication is available from the British Library

ISBN   978-1-108-83431-5   Hardback
ISBN   978-1-108-99488-0   Paperback

Contents

Acknowledgments

This project would not have been possible without the wonderful vision and generous support of the Polonsky Academy and the Polonsky family. I'd like to thank Tim Button, Geoffrey Hellman and Hartry Field for their detailed and generous feedback on parts of this book. I'd also like to thank Jared Warren, Tom Donaldson, Zeynep Soysal and Dan Waxman for much delightful philosophical argument over the internet. Going back further, I'd like to thank my dissertation advisors Warren Goldfarb, Peter Koellner, Ned Hall and Bernhard Nickel; my graduate school friends Jon Litland, Eylem Özaltun and James Shaw; and my former colleagues at the Polonsky Academy for Advanced Study and the ANU, especially Silvia Jonas, Casper Storm-Hanson, Olla Solomyak, Dave Chalmers, Daniel Nolan and Ray Briggs. Finally, I'd like to thank my dear husband Peter Gerdes for endless mathematical writing advice and lively philosophical debate, and my parents Edward and Li-shar Berry.

This book is dedicated to Robert Berry my beloved grandfather and first philosophical interlocutor, in memory of many long arguments. His stubborn curiosity, kindness and quest for rigor are an inspiration to me.

An online appendix with further details is availible at www.cambridge.org/PotentialistSetTheory

1   Introduction

Admittedly, the present state of affairs where we run up against the paradoxes is intolerable. Just think, the definitions and deductive methods which everyone learns, teaches, and uses in mathematics, the paragon of truth and certitude, lead to absurdities! If mathematical thinking is defective, where are we to find truth and certitude?

– Hilbert (Bencerraf and Putnam, 1983)

Philosophers have often used first-order logic to analyze mathematical and scientific claims. However, we seem to grasp a notion of logical possibility prior to, and independent from, our grasp of mathematical objects like set-theoretic models. Powerful reasons to accept this notion as an additional logical primitive have emerged (Boolos 1985; Etchemendy 1990; Gómez-Torrente 2000; Hanson 2006; Field 2008).

In this book, I'll make a case that philosophical analyses using (a natural generalization of) this notion of logical possibility can illuminate the philosophy of mathematics, metaphysics and philosophy of language. Much of this case will focus on pure mathematics and the philosophy of set theory. For example, I will show that formulating set theory in terms of logical possibility (along Potentialist lines suggested by Putnam and Hellman in response to the Burali-Forti paradox) yields a new and more appealing justification for one of the standard ZFC (Zermelo–Frankel with choice) axioms of set theory. This brings us closer to realizing the traditional hope of justifying mainstream mathematics from principles that seem clearly true.

However, we will see that using a primitive logical possibility operator can also help us develop a modestly neo-Carnapian philosophy of language. And philosophical analyses of scientific theories using the logical possibility operator can illuminatingly "factor" scientific claims into a logico-mathematical component and a remainder, in a way that reveals hidden heterogeneity in the role of mathematics in the sciences and clarifies debates over Quinean and post-Quinean indispensability arguments.

1.1   Mathematics as a Touchstone and the Centrality of Set Theory

Mathematical proofs provide a touchstone of clarity and convincingness which serves as an inspiration to philosophy and other disciplines. While it is possible to doubt the results of mainstream mathematical arguments (philosophers are capable of doubting anything), there's something striking about just how convincing mathematical proofs often are. Consider the standard argument that there are infinitely many primes. Even philosophers who deny that there are numbers (and hence think the argument as usually stated is unsound) are strongly tempted to say that we know *something like* the premises and that these proofs provide some kind of valuable amplification of this knowledge. The premises we use in informal mathematical reasoning have a combination of prima facie obviousness, power and generality, which makes them exemplary tools for expanding our knowledge and resolving disputes in cases where people's initial hunches disagree. It's no surprise that Leibniz[1] wished philosophers could resolve their disputes like mathematicians by saying "let us calculate" (or at least, "let us each look for a proof").

In many ways, set theory lies at the heart of modern mathematics, and it does powerful mathematical (not just philosophical) work as a foundation for the whole. So, one might hope that the set-theoretic foundations for mathematics would share the clarity and convincingness we hope for from mathematical arguments.

However, certain problems in the philosophical foundations of set theory raise worries. These concerns are more mathematical and specific to set theory than standard philosophical worries about, e.g., whether there are any abstract objects. And these concerns are more threatening to mathematical practice than philosophical doubts typically are, insofar as they raise doubts about whether the standard ZFC axioms of set theory are even logically consistent.

The development of set theory resolved a great many problems in analysis. It also provided a formal framework to allow interactions between various areas of mathematics – creating, as Hilbert (1926) famously observed, a kind of mathematical paradise. However, contradiction, in the form of Russell's paradox, threatened Hilbert's paradise.

This problem was almost, but not quite, solved by accepting the iterative hierarchy conception (also called the cumulative hierarchy conception) of sets and the standard ZFC first-order axioms for set theory. On the iterative hierarchy conception of set, we think about the sets as being formed in layers, with the sets at each layer containing only elements from prior layers. This lets us avoid the appearance that there should be a set of all sets that aren't members of themselves, and hence Russell's paradox. And it is hard to deny that the mathematical results which are currently stated in terms of this conception of set theory reflect genuine and important knowledge of some kind.

However, as we will see below, a few further problems arise, including a question of how to justify our acceptance of all the ZFC axioms. So, we may ask, is the price of

---

[1] See Chrisley and Begeer (2000: 14).

remaining in Cantor's set-theoretic paradise giving up the old ambition of founding mathematics on intrinsically obvious seeming principles?

One of the main projects of this book will be to answer this question in the negative. I will develop a unified understanding of set-theoretic talk, which vindicates our intuitive expectations regarding set theory – and demonstrates that all theorems of set theory (ZFC) are derivable from principles that seem clearly true (not to say indubitable). The approach I'll develop differs from standard Actualist set theory, in a few important ways.

1.2   Actualism and Its Discontents

On standard Actualist approaches to set theory, set theory studies abstract mathematical objects called "sets," which form an iterative hierarchy as evoked above. Apparent existence claims made by set theorists (like "there is a set which has no elements") are made true by the existence of corresponding objects, just like ordinary existence claims about cities or electrons or cars. Crudely speaking, three problems arise when we look at set theory from this familiar point of view (each of which I'll describe in much greater detail in Section 2.2).

First, there's a problem about our conception of the hierarchy of sets as a mathematical structure. We don't seem to have a precise conception of the intended structure of the iterative hierarchy of sets, in the way that we do *seem* to have a conception of the natural numbers. In particular, the height of the hierarchy of sets is left vague or mysterious. As the Burali-Forti paradox (Burali-Forti 1897) dramatizes, a certain naive conception of the hierarchy of sets (as containing ordinals corresponding to all ways some objects could be well ordered by some relation) is incoherent. And once this naive paradoxical conception of the height of the hierarchy of sets is rejected, we don't seem to have a precise idea about the intended height of the hierarchy of sets left over to replace it. It appears that, for any height that the hierarchy of sets could achieve, there could be a strictly larger structure, which extra layers of "sets" on top of the original hierarchy and fits with everything in our conception of the sets equally well. But it seems arbitrary to suppose that the hierarchy of sets happens to stop at any particular point.

Second, there's a worry about generality and the role of set theory as a foundation for all of mathematics. One might hope that set theory would be able to represent any mathematical structure one might want to study. The idea that set theory has this kind of generality is prima facie quite intuitive. But Actualist set theory is prima facie unable to represent the study of mathematical structures that are "too large." Thus, Actualism makes it hard to capture the intuition that "any possible structure" should, in some sense, be fair game for mathematical study, and hence treatment within set theory.

Third, there's a problem about intuitively justifying the standard ZFC axioms of set theory. As noted above, mathematical proofs can usually be reconstructed so as to derive their conclusion from premises that are prima facie extremely plausible (if not indubitable or impossible to philosophically or empirically cast doubt upon). So, one might hope that (once we understand set theory correctly), every claim provable from the ZFC axioms for set theory can be shown to follow from principles that seem clearly true in the way that foundational mathematical axioms often do.

However, philosophers have noted that the axiom of Replacement (one of the standard ZFC axioms) doesn't seem clearly true and deriving it from principles that do seem clearly true has proved challenging. For example, Hilary Putnam (2000) writes, "Quite frankly, I see no intuitive basis at all for . . . the axiom of Replacement. Better put, I do not see that a notion of set on which that axiom is clearly true has ever been explained." Instead, philosophers of mathematics and mathematicians have made do with less ambitious approaches to justification. For example, some mathematicians have invoked external justifications, like the failure of mathematicians to discover any contradictions during over a century of work with ZFC. Others have shown that the axiom of Replacement follows from certain powerful and plausible (if not clearly consistent or true) principles that also imply many of the other axioms of set theory. But insofar as the latter powerful conceptions aren't clearly consistent, showing this doesn't suffice to show Replacement follows from clearly true premises. Nor does simply combining Replacement itself with the bare-bones conception of the intended width of an iterative hierarchy of sets evoked above yield something that seems obviously consistent. This state of affairs can feel unsatisfying.

1.3   Potentialism and the Justification Problem

In response to the first two problems above, philosophers like Putnam, Parsons, Hellman and Linnebo (Putnam 1967; Parsons 1977; Hellman 1994; Linnebo 2018a) have proposed that we should reject[2] Actualism about set theory in favor of a different approach: Potentialism. The key idea behind Potentialism is that, rather than taking set theory to be the study of a single hierarchy of sets which stops at some particular point (as the Actualist does), we should instead interpret set theorists as making modal claims about what hierarchy-of-sets-like structures are possible and how such structures could (in some sense) be extended.

As we will see in more detail in Chapter 2, switching to a Potentialist understanding of set theory solves the first problem for Actualism noted above. The Potentialist avoids postulating an arbitrary (or indeterminate) height for the hierarchy of sets,[3] and Potentialism also plausibly solves the second problem above, by honoring the intuition that any possible mathematical structure can be studied within set theory.

---

[2] Strictly speaking Putnam proposes Actualism and Potentialism are (in some sense), two perspectives on the same thing.

[3] At least Potentialists like Hellman (1996), Linnebo (2018a) and Studd (2019) avoid positing such an arbitrary stopping point for the sets. Putnam's view, on which Actualist set theory and Potentialist set theory are (somehow) two perspectives on the same thing, does not let us avoid this problem in any obvious way.

However, the problem of justifying the axiom of Replacement from premises that seem clearly true remains. Contemporary Potentialists can, and generally do, prove that (the Potentialist translations of) every theorem of ZFC can be derived from certain intuitive assumptions about logical possibility, or some other such modal notion. However, these proofs all use principles of modal logic that aren't (and aren't claimed to) be clearly true in the way invoked by Putnam. The existing Potentialist literature has shown that Potentialism is *no worse off than Actualism* with regard to the problem of justifying Replacement that Putnam raises.[4] However, neither Potentialists nor Actualists have put forward a solution to this problem.

1.4 Outline

In this book, I will attempt to solve the above problem of justifying the axiom of Replacement from principles that seem clearly true (or at least improve on existing solution attempts) and clarify the foundations of set theory.

In Part I, I will argue that we should indeed be Potentialists about set theory for essentially the reasons indicated above, and then review major existing formulations of Potentialism about set theory and some problems for each. I'll discuss and contrast two major existing versions of Potentialist set theory: the Putnamian approach developed by Putnam and Hellman which I will largely follow, and an alternative Parsonian approach recently explored by Linnebo and Studd, which appeals to a notion of interpretational possibility, rather than metaphysical or logical possibility.

I will develop and advocate a particular form of Potentialist set theory. Although this approach largely blends and simplifies ideas from Putnam and Hellman, it has the distinctive feature of replacing claims that "quantify-in" to the diamond of logical possibility (and thereby talk about what's logically possible *for objects*) with claims about what's logically possible *given certain structural facts*, expressed using a new piece of logical vocabulary I'll call the conditional logical possibility operator ($\Diamond_{...}$). Cashing out Potentialist set theory in these terms lets us avoid certain philosophical controversies,[5] as well as practically helping us state axioms that can be easily grasped and recognized as saying something clearly true.

In Chapter 2 I will discuss Actualist approaches to set theory and expand on the worries for them noted above. In Chapter 3 I'll discuss how adopting some Potentialist approach to set theory promises to solve these worries and review existing forms of the Putnamian style of Potentialism. I will defend Hellman's use of a notion of logical possibility to cash out Potentialist set theory but note that controversies over quantified modal logic raise some problems for using his version of Potentialism in our foundational project.

---

[4] Existing potentialists (Hellman 1994; Linnebo 2018a; Studd 2019) have generally adopted some version of a Potentialist translation of Replacement as an axiom (schema). For while these Potentialist translations are not clearly true, they are (we will see) as attractive as corresponding instances of the Replacement schema understood actualistically.

[5] See Section 3.3.1.

In Chapter 4 I'll introduce my preferred style of Potentialist paraphrase and the key notion of conditional (structure-preserving) logical possibility I'll use to give these paraphrases. Finally, in Chapter 5 I'll contrast the above approach to Potentialist set theory with those advocated by Linnebo and Studd, major proponents of an alternate "Parsonian" school of Potentialist set theory.

In Part II I will turn to the core mathematical project of this book: justifying the ZFC axioms. I'll propose general purpose axioms for reasoning about conditional logical possibility which (I claim!) seem clearly true in the way our foundational project requires. Then I will show that these axioms justify our use of normal first-order reasoning for set-theoretic claims (i.e., claims in the first-order language of set theory) even when those claims are understood potentialistically. Specifically, if we let $\phi^\Diamond$ stand for the Potentialist translation of a set-theoretic claim ϕ, let $\vdash_{FOL}$ be provability in first-order logic and $\vdash$ be provability in the formal system proposed in this book, we can show the following.

Theorem 1.1 (Logical Closure of Translation). *Suppose Φ, Ψ are sentences in the language of set theory and $\Phi \vdash_{FOL} \Psi$ then $\Phi^\Diamond \vdash \Psi^\Diamond$.*

With this theorem in mind, all that's needed to justify normal mathematical practice is to demonstrate that if ϕ is an axiom of ZFC then $\vdash \phi^\Diamond$ holds. A key idea here will be to use certain *non-interference* intuitions to justify the (Potentialist translation of) the axiom of Replacement, rather than simply taking the latter as an axiom, as current Potentialists tend to do. Putting these pieces together we can conclude that for all set theoretic sentences ϕ:

$$\text{ZFC} \underset{\text{FOL}}{\vdash} \phi \text{ then } \vdash \phi^\Diamond$$

That is, reasoning in ZFC as if one were talking about an Actualist hierarchy of sets is harmless. If one can prove that ϕ in ZFC then the Potentialist translation of ϕ (written $\phi^\Diamond$ above) is (true and indeed) provable in my formal system.

Note that since I choose axioms of reasoning about conditional logical possibility which are attractive for general use rather than ones that directly mirror Actualist ZFC set theory (as other Potentialists have done in proving versions of the theorem above), it's not at all obvious whether the reverse direction of the above conditional, i.e., "If $\vdash \phi^\Diamond$ then ZFC $\vdash_{FOL} \phi$" is true. In principle, there is some hope that the modal axioms I propose (or, more plausibly, further principles about conditional logical possibility that seem equally clearly true) will let one vindicate new axioms for set theory, going beyond the ZFC axioms.

Finally, in Part III of the book, I'll turn to larger philosophical questions. In Chapter 10, I consider two ways my story about set theory can fit into a larger philosophical picture of mathematics and its applications: a Nominalist approach and the weakly neo-Carnapian approach I ultimately favor.

In Chapters 11–14, I'll discuss the Nominalist approach to non-set theoretic mathematical objects and Indispensability arguments. I'll argue that adding some cheap tricks to the above paraphrase strategy lets the Nominalist answer certain classic Quinean and Explanatory indispensability arguments. However, I'll suggest that the mathematical Nominalist *may* face serious and under-discussed worries about reference and grounding.

In Chapters 15 and 16, I'll explain the weakly neo-Carnapian approach to non-set theoretic mathematical objects I favor, and argue that adopting it helps solve or avoid these reference and grounding problems and has certain other advantages (while retaining many benefits of Nominalism). The resulting view is a kind of neo-Carnapianism realism about mathematical objects, which drops Carnap's radical anti-metaphysical ambitions but keeps mathematicians' freedom to talk in terms of arbitrary logically coherent pure mathematical structures.

Finally, in Chapters 17–19, I'll discuss how the overall picture of mathematics developed in this book relates to traditional questions about Logicism, Structuralism and human access to facts about objective proof-transcendent mathematical facts.

1.5   Caveats and Clarifications

Let me finish this introduction with some quick caveats about the nature and aim of my project.

First, I don't claim set theorists should literally rewrite set theory textbooks in Potentialist terms. Mathematicians' current practice of (making arguments which can be reconstructed as) proving things in first-order logic from the ZFC axioms is fine. And doing something like logical deduction from purely first-order axioms may be unavoidably easier (for minds like ours) than thinking about the elaborate modal claims that figure in Potentialist set theory. If one thinks about typical set theoretic talk as abbreviating Potentialist claims, then the main result of Part II shows that it's unnecessary to unpack this abbreviation in most mathematical contexts.

However, I *am* suggesting Potentialism reflects what people should say when we think about set theory in many philosophical and foundational mathematical contexts. Replacing Actualist set-theoretic claims with their Potentialist paraphrases solves various intuitive puzzles and makes sense of things that we normally want to say about set theory.[6]

---

[6] In this proposal I somewhat mirror Hellman's response (Hellman 1998) to Burgess and Rosen's dilemma (Burgess and Rosen 1997). Burgess and Rosen argue that nominalistic paraphrases must be intended as either a hermeneutic theory of what scientists mean or a revolutionary theory of what they should say, but typical Nominalist paraphrases don't seem adequately supported by scientific motivations for either as they wouldn't be accepted to linguistics or physics journals.

One response to this would be to say that nominalistic paraphrases reflect what we should say in *philosophical* contexts, and this differs from what we should or do say in any scientific context. Hellman points out that one can appeal to useful divisions of labor within the sciences to motivate such a distinction. For example, a physicist who hypothesizes that heat is molecular motion (and regiments physical theories accordingly) isn't thereby making a revolutionary proposal about what higher-level scientists (biologists or ecologists) should say or a hermeneutic proposal about what they currently mean. So the untroubled friend of metaphysics can think about ontology as its own discipline, with its own level of analysis and corresponding explanatory work this analysis is invoked to perform. A Nominalist of this stripe might say: metaphysics is to physics as physics is to biology and ecology. That's why good proposals about what we

Arguably, this book's project of developing Potentialist foundations for set theory is analogous to the familiar project of providing a set-theoretic foundation for analysis. Our naive reasoning about certain concepts (limits in one case, the height of an iterative hierarchy of sets that "goes all the way up" in the other) turns out to lead to paradox. So, it is desirable to find a different way of thinking about relevant mathematical claims which will let us capture their intuitive significance and interest, while blocking paradoxical inferences.

Second, the Potentialist understanding of pure set theory advocated in Parts I and II of this book is compatible with a range of different views about how to understand other areas of mathematics. I hope my version of Potentialism will be compelling even to readers who find both Nominalism and the neo-Carnapian realism about mathematical objects (outside set theory) I advocate in Part III unacceptable.

Third, I aim to provide a foundation for Potentialist set theory which rests entirely on intuitively compelling principles that are subject matter neutral and constrain the behavior of all objects (c.f., Frege's characterization of logic in Frege (1980)). Thus, in a sense I'm defending a kind of Logicism about set theory. But I don't mean to claim that my foundational principles are analytic, cognitively trivial, or impossible for any rational being to doubt. I merely claim they're clearly true in the sense evoked by Putnam above.[7] I also don't mean to suggest that facts about conditional logical possibility discussed in this book constitute some kind of metaphysical free lunch.[8]

Fourth, we must distinguish the foundational project in this book from a less ambitious justificatory project. Actualist philosophers have sometimes aimed to find a unified conception of set theory from which all the various ZFC axioms clearly follow – without worrying whether this conception itself is clearly coherent. This project can be valuable in various ways, e.g., in showing the naturalness and appeal of certain mathematical hypotheses (like proposed large cardinal axioms) which also follow from the relevant conception. However, finding such a unified conception doesn't suffice for my foundational project. For if the unifying conception isn't clearly *consistent* then, surely, it isn't clearly *true* (even on a view which allows

---

should start to say in philosophy journals can differ radically from what physics journals would or should publish. Perhaps Sider's distinction between metaphysical semantics and linguistic semantics discussed in Section 11.4.2 suggests a similar line of response (Sider 2011).

   However, the motivations I urge for Potentialist set theory are closer to those for foundational projects within mathematics than the explicitly philosophical motivations Hellman and Sider reference. Thus, I think the Potentialist paraphrases I advocate might be accepted by extreme naturalist readers, who would reject the above suggestion that philosophy or metaphysics could provide a legitimate further level of analysis beneath the sciences. Also note that the motivations for Potentialist set theory I press in this text aren't among the specific philosophical motivations for Nominalist formalizations of mathematics which Burgess and Rosen (1997) criticize.

[7] I take the axiom of choice to be prima facie clearly true, despite the fact that it can be doubted on grounds like the Banach–Tarski paradox. But readers who find Choice less immediately appealing can read this as a claim to justify Replacement from principles "as prima facie obvious as the other axioms of *ZF* set theory" instead.

[8] I take accepting a primitive modal notion of (conditional) logical possibility to be a significant, but warranted, addition to our fundamental ideology.

mathematicians to introduce arbitrary coherent structures). So, we haven't succeeded in justifying all theorems of ZFC set theory from premises that seem clearly true.

Finally, the set-theoretic foundational project of this book also differs from a *more* ambitious project, along the following lines. Philosophers sometimes seek the most metaphysically joint-carving successor to the naive concept of sets which generates Russell's paradox (something which might, e.g., be hoped to connect intimately with the true answer to the liar paradox). So, for example, you could ask whether the iterative hierarchy conception of sets is remotely on the right track, or whether the "best" successor to naive set theory is something like Quine's New Foundations instead.

I think this more philosophically speculative project is legitimate, but not required for the foundational justificatory project attempted in this book. I will try to show that Potentialist translations which attractively explicate theorems of current mainstream set theory follow from principles that seem clearly true. But I won't take a position on what the most metaphysically illuminating successor to naive set theory is.

Part I

2 Actualist Set Theory

In this chapter, I will discuss the traditional, Actualist, approach to set theory. I will review how the Actualist faces problems articulating a categorical conception of the intended height of the hierarchy of sets (despite the existence of certain categoricity and quasi-categoricity theorems). I will then discuss how the Actualist faces problems justifying the axiom of Replacement from principles that seem clearly true.

2.1 Actualist Set Theory and the Iterative Hierarchy Conception

On a straightforward Actualist approach to set theory, there are abstract objects called "the sets," much as there are abstract objects called "the natural numbers." And we can ask: what sets exist? And what kind of structure do the sets have under the relation of membership?

Naively one might want to say that, for any formula $\phi(x)$, there is a set whose elements are exactly those objects that satisfy ϕ. But, as Bertrand Russell famously showed, this leads to paradox as there must be a set whose elements are exactly the sets which aren't members of themselves.

The (widely embraced) iterative hierarchy conception of the sets solves this problem by suggesting a different picture of what sets exist. On this picture, we think about the sets as forming layers, with sets at a given layer in the hierarchy only being able to have elements that are available at previous layers. Each layer contains "all possible sets" of elements given at prior layers, and no two sets have exactly the same elements.[1] Talk about the height of such a hierarchy of sets refers to the "number" of layers, while talk about its width refers to how many sets are introduced at each stage.

One can spell out this idea of a full-width iterative hierarchy as follows.

---

[1] Note that there's been some discussion about whether extensionality follows from the iterative concept of sets or is something separate. But the worries I raise for Actualists won't depend on the idea that our conception of the hierarchy of sets must be "unified" in this strong sense. The question I will be pressing in Section 2.4 is merely whether we have a coherent conception of the hierarchy of sets (once the incoherence of our naive conception of the hierarchy of sets is recognized) that even seems to pick out a unique structure, not whether that conception is unified in the strong sense evoked above.

Definition 2.1 (Iterative Hierarchy – Full Width (IHW)). A full-width iterative hierarchy (IHW) is a structure consisting of:

- a well-ordered series of levels; and
- a collection of sets "available at" these levels, such that:

  – at each level, there are sets corresponding to "all possible ways of choosing" some sets available at lower levels (note that this can be stated straightforwardly in second-order logic)
  – sets x and y are identical iff they have exactly the same members (extensionality).

One can think of IHW as specifying a structure for initial segments of the hierarchy of sets. If we adopt the idea of a hierarchy of sets, then the principles above specify an intended width for this structure. One can (clearly) formalize the above claim using second-order logic, and I'll refer to the resulting theory as IHW_2.[2]

In contrast the ideas evoked above do not pick out a unique intended height for the hierarchy of sets.[3] Indeed, as we will now review, there are important reasons for doubting that we have any coherent and adequate conception of "absolute infinity," the supposed height of the hierarchy of sets. And the version of Potentialism I favor will wind up denying that there is, strictly speaking, a hierarchy of sets (hence anything for mathematical talk of "the height of the hierarchy of sets" to refer to[4]).

2.2   A Burali-Forti Problem

The problem for actualist set theory is not simply that it might be impossible to define the notion of absolute infinity in other terms. After all, every theory will have to take some notions as primitive.

Instead, we find ourselves in the following situation:

- Our naive conception of absolute infinity (the height of the Actualist hierarchy of sets) turns out to be incoherent, not just unanalyzable.
- And, once we reject this naive conception, there's no obvious fallback conception that *even appears* to specify a unique height for the hierarchy of sets in a principled and sufficiently clearly consistent way.

Specifically, a very common intuitive conception of the hierarchy of sets says that the hierarchy of sets goes "all the way up" – so no restrictive ideas of where it stops are

---

[2] However, my preferred approach will reject the formalization in second-order logic in favor of one $IHW_\Diamond$ using only the conditional logical possibility operator $\Diamond_{...}$ introduced in Chapter 3. I'll understand IHW loosely to be compatible both with a Boolos style two-sorted conception and the standard cumulative hierarchy.

[3] We could add the principle that there is no last stage, as Boolos (1971) does. But since there are many different logically possible well-orderings which do not have a last element, e.g., ω, $\omega + \omega$, etc., this does still not give us a unique intended height.

[4] Instead we will analyze set-theoretic talk as expressing Potentialist claims about logical possibility, and extendability.

needed to understand its behavior. However, if the sets really do go "all the way up" in this sense, then it would seem that they should satisfy the naive height principle:

Naive Height Principle For any way some things are well-ordered by some relation R, there is an ordinal corresponding to it.

But, for example, the ordinals themselves are well-ordered, and there is no ordinal corresponding to this well-ordering, i.e., there is no ordinal which has the same order-type as the class of all ordinals. Thus (it would seem), the naive height ordering principle above can't be correct, and it seems arbitrary to say that the hierarchy of sets just stops somewhere if a suitable stopping point is not pinned down by something in our conception of the hierarchy of sets.

To clarify this worry, note that I'm not suggesting the Actualist must think the hierarchy of sets "must stop somewhere," in the sense that they must say there's a largest ordinal. There's no problem about saying that for every set/ordinal x there's a strictly larger set/ordinal y. Nor do I mean to suggest that there could somehow be "sets beyond all the sets," or that there's something wrong with taking various concepts used in articulating a conception of the hierarchy of sets as primitive (it's hard to see how one could avoid doing the latter!).

Rather, the problem is that the Actualist takes there to be some plurality of objects (the sets) forming an iterative hierarchy structure i.e., satisfying the description of the intended *width* of the hierarchy of sets above. But the following modal intuition seems appealing: for any plurality of objects satisfying the conception of an iterative hierarchy above (i.e., for any model of IHW), it would be *in some sense* (e.g., conceptually, logically or combinatorially if not metaphysically) possible for there to be a strictly lager model of IHW which, in effect, adds a new stage above all the ordinals within the original structure together with a corresponding layer of classes.[5] And, worryingly, it seems that the resulting structure generated would answer everything in our conception of the sets as well as the original structure did. For, once we've rejected the naive conception of the intended height of the hierarchy of sets above as inconsistent, we don't seem to have anything that even pretends to pick out a unique height.

Thus, the Actualist seems forced to say that the plurality of existing sets just happens to instantiate one possible structure. The hierarchy of sets just happens to have some particular height, although nothing in our conception of the sets rules out epistemic possibilities where the hierarchy of sets is taller.

But saying that the hierarchy of sets just happens to stop at a certain point seems to violate intuitive principles of metaphysical parsimony. It seems to require acknowledging an extra – otherwise entirely unmotivated – joint in reality, namely the height of the hierarchy of sets. One might also worry about the epistemology of this stopping point: why should we think set theorists' reasoning about large cardinals etc., correctly reflects this brute fact about where the hierarchy of sets happens to stop?

---

[5] I won't say more about how to spell out the informal notion of possibility being invoked here now, but each version of Potentialist set theory discussed below (mine included) brings with it a candidate modal notion.

The simplest response to this problem might be to find some other restrictive characterization of the sets (in particular, some other characterization of the intended height of the hierarchy of sets).[6] However, there's no obvious fallback/replacement conception that even seems to pick out a unique structure. It's not clear that *any* precise intuitive conception of the intended height of the sets remains once the paradoxical well-ordering principle above is retracted. As Shapiro and Wright (2006) put it, all our reasons for thinking that sets exist in the first place appear to suggest that, for any given height which an actual mathematical structure could have, the sets should continue up past this height.

Moreover, the sets lose a substantial aspect of their appeal as a mathematical foundation if we can't capture all talk of coherent mathematical structures within set theory, i.e., via quantification over the sets or some set model that's at least isomorphic to the relevant mathematical structure. However, it is (at best) unclear whether we can do this if we accept Actualism and say that the hierarchy of sets doesn't "go all the way up" in the sense indicated above. Of course, by Gödel's completeness theorem for first-order logic, any consistent collection of first-order axioms will have a model. However, our conceptions of mathematical structures (like, famously, the natural numbers) can include non-first-order notions. So, the completeness theorem doesn't guarantee that our conceptions of these structures will have "intended" models in the hierarchy of sets (i.e., models which treat their non-first-order vocabulary standardly). One might further press this objection by arguing as follows. If there were an Actualist hierarchy of sets we could refer to, then we could also uniquely describe the possible structure which we would get by adding a single layer of classes to this hierarchy of sets. This structure is a legitimate topic for mathematical investigation, and yet this structure is not instantiated anywhere within the hierarchy of sets.[7]

Note that, if some Actualist claimed to have a suitably primitive and seemingly precise notion of absolute infinity, they wouldn't face the arbitrariness worry I'm pressing. They could appeal to this notion of absolute infinity to specify the height of the hierarchy of sets. However, even though people do use the term "absolute infinity," this seems to be little more than a name for whatever height the hierarchy of sets has. They don't claim to have a concept that seems capable of picking out a precise intended height without deference to prior facts about however tall a hierarchy of sets there happens to be. Arbitrariness troubles arise because we start out with the seemingly precise naive conception of the intended height of the hierarchy of sets, and no other seemingly precise notion appears to fill the gap once this naive conception is rejected as paradoxical.[8]

---

[6] Note that the axioms of ZFC and even ZFC_2 don't suffice to categorically determine the height of the hierarchy of sets.

[7] See Hellman (1994) for a version of this generality worry.

[8] That is, I take it most Actualists would agree that we don't even *seem* to have an independent precise (primitive or otherwise) conception of the intended height of the hierarchy of sets in the way that (many

Now, we could avoid the above worry about arbitrariness while securing a definite height for the hierarchy of sets by simply *adding* some new idea about height to our current conception of the hierarchy of sets. For example, it might seem natural to say that the sets are the shortest possible structure satisfying ZFC_2 (i.e., the hierarchy of sets, so to speak, stops below the first inaccessible). This proposal is natural as it mirrors how we pick out a unique structure for the natural numbers by saying that the numbers are "as short as can be" while being closed under successor. However, making this kind of height-minimizing stipulation seems to fit badly with actual mathematicians' interest in large cardinals (which require the set-theoretic hierarchy to extend far beyond the shortest model of ZFC). And, more generally, stipulating any height for the hierarchy of sets does nothing to help with the secondary worry above, that Actualists shortchange the intended generality of set theory.

2.3   Categoricity and Quasicategoricity Arguments

2.3.1   McGee and Appeal to Ur-elements

With this worry about stating a precise conception of the hierarchy of sets (and avoiding arbitrariness) in place, let me quickly explain why two categoricity theorems which might seem to help the Actualist don't help her.

In "How We Learn Mathematical Language," McGee (1997) advocates a conception of an iterative hierarchy of sets with ur-elements, and proves a "quasi-categoricity" theorem about it, which might seem to help the Actualist address the arbitrariness challenge posed above.

However, I will argue that this is an illusion. Although McGee's characterization of a hierarchy of sets solves the problem he is concerned with in that paper (addressing a certain kind of referential skepticism), it does not make the height of the Actualist hierarchy of sets look any less arbitrary.

McGee (1997) defends realist claims that we can secure definite reference to the hierarchy of sets up to isomorphism (and thereby justify our presumption that all questions in the language of set theory have definite right answers) from a reference skeptical challenge.

Specifically, he proposes an account of how creatures like us could count as having a definite conception of the sets up to isomorphism, given the presumption that we can secure definite realist reference for other kinds of vocabulary, and (it will be important to note) that we are somehow able to quantify over everything (sets included).

---

would say) we do *seem* to have a conception of the intended width of the hierarchy of the sets or what second-order collections or pluralities there are supposed to be. An Actualist who (unlike all the Actualists I've encountered) did claim to grasp a primitive notion of absolute infinity that picked out a precise structure in this way would not face the arbitrariness problem above. See Section 2.5.2 for much more detail regarding this distinction.

McGee explains how we can secure (the effect of) definite reference to second-order quantification and thus uniquely describe the intended width of the hierarchy of sets, via a story about schemas which I won't summarize here. Then he suggests that we can pin down the intended height of the hierarchy of sets by considering a conception of a hierarchy of sets *with ur-elements*.

The idea of set theory with ur-elements is simply to allow sets to have elements that aren't sets. One keeps the core idea of an iterative hierarchy of sets described above (with each layer containing "all possible subsets" from the lower layers), but takes the lowest level of the hierarchy of sets to include sets corresponding to all ways of choosing from among all the objects that aren't sets (e.g., elephants, billiard balls, electrons, marriages and the like), rather than just the empty set. Note that the hierarchy of sets with ur-elements includes all pure sets. Thus, uniquely pinning down a hierarchy of sets with ur-elements would suffice to pin down a hierarchy of pure sets as well.

The Ur-element Set Axiom follows from the statement above, and says that there's a set which contains, as elements, all the objects that aren't sets:

Ur-element Set Axiom (U) $(\exists x)(Set(x) \wedge (\forall y)(\neg Set(y) \rightarrow y \in x))$

McGee shows that we can (in a sense) pin down the intended height of this hierarchy of sets with ur-elements if we accept the axiom above.

Specifically, McGee proves that $ZFC_2 + U$ (the result of adding the above ur-element principle to second-order ZFC set theory) has a property which he calls "quasi-categoricity."[9] Given any single choice of a total domain (what you are quantifying when you quantify over everything *including the sets*) there cannot be two non-isomorphic (with respect to $\in$) interpretations of set theory which simulaneously: choose "sets" from within this domain, take quantifiers to range over this whole domain and make McGee's $ZFC_2 + U$ come out true (while interpreting all logical vocabulary standardly).

McGee's theorem ensures that we couldn't have a single universe containing both a hierarchy of red sets and a hierarchy of blue sets, such that both hierarchies satisfy the constraints imposed by $ZFC_2 + U$ on their relationship to the total universe (red sets and blue sets included). So, it does the job McGee wants: answering skeptical challenges about definite reference to the hierarchy of sets (up to isomorphism), on behalf of a Platonist who presumes that there's an Actualist hierarchy of sets and grants that we can somehow unproblematically quantify over everything (sets included).

However, this theorem does nothing to address the objection to Actualism raised at the beginning of this chapter: that Actualists seem committed to an additional and arbitrary joint in reality – a point where the hierarchy of sets just happens to stop.

For McGee's theorem does not imply that we have any beliefs which logically necessitate (and thereby make non-arbitrary) facts about where the hierarchy of sets

---

[9] One might worry about the above axiom on the basis of Uzquiano's proof that McGee's axioms for set theory with ur-elements are incompatible with certain axioms of mereology (Uzquiano 1996), but I leave this question aside as the concerns I will be raising are unrelated.

happens to stop. As McGee himself points out, the conception of sets he articulates is **not** categorical; the beliefs about the sets which he invokes are compatible with many different possibilities about how large the total universe of sets is.

Indeed, it's crucial to notice, McGee's theorem doesn't even show that $ZFC_2 + U$ is *quasi*-categorical in the following (to my mind, more natural) sense of the term. It doesn't show that, fixing the facts about what non-set objects there are, any hierarchy of sets satisfying $ZFC_2 + U$ must have a certain unique structure. Indeed, given certain popular assumptions you can always[10] take one possible scenario containing a hierarchy of sets satisfying $ZFC_2 + U$ within a total universe of a certain size, add some sets to the top of this hierarchy, and therefore to the universe (without changing any facts about the non-sets), and get another possible scenario satisfying $ZFC_2 + U$.

Thus, McGee's theorem doesn't pin down a unique intended structure for the hierarchy of sets or abolish arbitrariness by explaining why the hierarchy of sets stops at some particular point. It just shows that you couldn't have two non-isomorphic hierarchies of sets satisfying the above conception within the same universe.

One could use McGee's conception of sets with ur-elements in a slightly different way which *would* block the arbitrariness worries for Actualism I've pressed above, as follows. Assume that our use of non-mathematical vocabulary to pins down the intended interpretation of certain non-mathematical kind terms. We could specify the intended height of the hierarchy of sets by saying that (in effect) the hierarchy of sets stops *as soon as it can* while satisfying $ZFC_2 + U$.

Unfortunately, however, this proposal faces the same worries about making the hierarchy of sets too small which arose for the idea that we could just pick a restrictive conception of the sets in Section 2.2. It also suggests the height of the hierarchy of sets might be contingent and that the result of physical and metaphysical investigation into how many non-mathematical objects there are should have bearing on facts about pure set theory in a way that seems potentially uncomfortable.

2.3.2 Martin

Similarly, Martin's categoricity theorem about set theory in Martin (2001) might at first sound like it helps the Actualist with the arbitrariness/lack of a definite conception worry, but actually does not. Indeed, Martin seems to positively endorse a version of this worry (Martin n.d.).[11]

---

[10] For instance, if we presume the existence of unboundedly many inaccessibles, as is often thought plausible, we are guaranteed multiple models of $ZFC_2 + U$ with a particular collection of ur-elements.

[11] There he distinguishes five ingredients in our conception of the hierarchy of sets as follows.
  The modern, iterative concept has four important components:

  1. the concept of the natural numbers;
  2. the concept of sets of x s;
  3. the concept of transfinite iteration;
  4. the concept of absolute infinity.

  Perhaps we should include the concept of Extensionality as Component (0).

Martin (2001) argues against plenitudinous anti-objectivist "multiverse" approaches to set theory (like Hamkins 2012) on which certain set-theoretic claims Φ are not determinately true or false for the following reason:

Multiverse Idea: The platonic realm of mathematical objects includes many different (non-isomorphic) hierarchies of sets. There's no unique intended V, even up to width. Rather each hierarchy V in the multiverse is expanded by some larger one which adds, e.g., a "missing" subset of the natural numbers V. (So, we might note, none of these hierarchies can answer our conception IHW of the width of the hierarchy of sets above.) Some of these Vs make Φ true and others make Φ false. And all of them are (absent specific mathematical choice to "work in" a particular hierarchy of sets) equally intended.

Martin argues against this multiverse proposal by noting that if we accept a certain conception of the hierarchy of sets (including the principles below), we can derive that there could not be two different hierarchies of sets ("the red sets" and "the blue sets"):

- A "uniqueness" principle: all sets are extensional. That is, if there are two distinct sets x and y (even in two different putative hierarchies!), then there must be some object which is an element of x but not y or vice versa. Thus, for example, there can be only one set *Mars, Venus*.
- A conception of the hierarchy of sets, including (among other more familiar elements) the following height closure principle: if a set exists, then any hierarchy of sets containing the elements of that set must contain the set itself.

Martin points out that it follows from the principles above, essentially by induction, that there can't be two different hierarchies of sets. Any two putative hierarchies of sets satisfying the conditions above must agree on their ur-elements, and then on their first layer and their second layer etc.

One can call this a categoricity result. But it doesn't answer our worry about arbitrary stopping points. For it doesn't imply that it's logically or metaphysically necessary that any collections of objects which satisfy the above conception of sets must have a certain (unique) structure. Rather, it merely shows that there can't be two distinct *actual* set-theoretic hierarchies. For example, Martin's argument doesn't rule out the possibility that there could be some description of an ordinal ϕ_κ, such that it would be logically possible to have a structure satisfying our conception of the sets containing an ordinal satisfying ϕ_κ but also logically possible to have such a structure which didn't contain any ordinal

And then he expresses the following reservations about whether we have a definite coherent notion of absolute infinity:

...so I am using the term "absolute infinity" for the concept that is the fourth component of the concept of set. One can argue that the concept is categorical, and that any two instantiations of the concept of set (of the concept of an absolutely infinite iteration of the sets of x's operation) have to be isomorphic. But it is hard to see how there could be a full informal axiomatization of the concept of set. There are also worries about the coherence of the concept. People worry, e.g., that if the universe of sets can be regarded as a "completed " totality, then the cumulative set hierarchy should go even further. Such worries are one of the reasons for the currently popular doubts that it is possible to quantify over absolutely everything. I am also dubious about the notion of absolute infinity, but this does not make me question quantification over everything. (Martin n.d.)

satisfying ϕ_κ. It merely shows that we couldn't have two actual hierarchies of sets satisfying Martin's assumptions, one of which contains ϕ_κ while the other does not.

2.4   A Problem Justifying Replacement

In addition to the worry above (about whether we have a coherent conception of the intended height of the hierarchy of sets), set-theoretic Actualists also face a problem about justifying the axiom schema of Replacement. They must make it plausible that whatever unique height (and hence structure) they think the hierarchy of sets has, allows it to satisfy Replacement.

Informally, the axiom schema of Replacement tells us that the image of any set under a definable (with parameters) function is also a set. More formally, let ϕ be any formula in the language of first-order set theory whose free variables are among $x, y, I, w_1, \ldots, w_n$. We can think of the formula $\phi(x, y)$ (and choice of parameters) as specifying a definable function taking x to the unique y such that $\phi(x, y)$. Then the instance of axiom schema of Replacement for this formula ϕ says the following:

$$\forall w_1 \forall w_2 \ldots \forall w_n (\forall a [\, \forall x (x \in a \rightarrow (\exists! y) \phi(x, y, w_1, \ldots, w_n))])$$
$$\leftrightarrow \exists b \forall x ((x \in a \rightarrow \exists y (\, y \in b \wedge \phi(x, y, w_1, \ldots, w_n))))]$$

So, Replacement says that whenever some first-order formula defines a function on a set a, i.e., associates each element x of a with a unique y, there is a set b equal to the image of a under this function. In other words, the hierarchy of sets extends far enough up that all the elements in the image of a can be collected together.

As Boolos (1971) points out, the axiom of Replacement imposes a kind of closure condition on the height of the hierarchy of sets, which doesn't obviously follow from the iterative hierarchy conception of the sets above, even if we add the claim that there is no last stage. For consider $V_{\omega + \omega}$. This structure satisfies both of the assumptions in IHW plus the extra claim that there isn't a last layer. However, it doesn't satisfy Replacement, since you could take the set ω (formed at layer $V_{\omega+1}$) and write down a function ϕ which associates 1 with $\omega + 1$, 2, with $\omega + 2$ etc. Then, for each x in ω, there's a y in $V_{\omega + \omega}$ satisfying $\phi(x, y)$. But there isn't any set b in $V_{\omega + \omega}$ which collects together the image of every member of ω. That set b is only formed at a $V_{\omega + \omega + 1}$. This raises a worry about how to justify Replacement, and (indeed) whether mathematicians are justified in using it at all.

So (even if we take for granted that there are objects satisfying the iterative hierarchy conception of sets), if we want to justify use of the ZFC axioms, a question remains about how to justify the axiom of Replacement.

There has been much interest and sympathy with this worry in the subsequent literature. As mentioned in the introduction, Hilary Putnam (2000) writes, "Quite frankly, I see no intuitive basis at all for ... the axiom of Replacement. Better put, I do not see that a notion of set on which that axiom is clearly true has ever been explained."

More recently, in a discussion of the history of set theory, Michael Potter remarks that, "it is striking, given how powerful an extension of the theory Replacement represents, how thin the justifications for its introduction were,"[12] and then reports of our present situation that, "In the case of Replacement there is, it is true, no widespread concern that it might be, like Basic Law V, inconsistent, but it is not at all uncommon to find expressed, if not by mathematicians themselves then by mathematically trained philosophers, the view that, insofar as it can be regarded as an axiom of infinity, it does indeed, as von Neumann . . . said, 'go a bit too far'" (Potter 2004).

To my knowledge, four main (Actualist) strategies for justifying Replacement are currently popular. First, one can try to justify the axiom of Replacement "extrinsically" in the way we often justify a scientific hypothesis, by appeal to its fruitful consequences, arguing it helps prove many things we independently have reason to believe and hasn't yet been used to derive contradiction or consequences we think are wrong. See Koellner (2009) on Gödel's proposal:

Even disregarding the intrinsic necessity of some new axiom, and even in case it had no intrinsic necessity at all, a decision about its truth is possible also in another way, namely, inductively by studying its "success", that is, its fruitfulness in consequences and in particular in "verifiable" consequences, i.e., consequences demonstrable without the new axiom, whose proofs by means of the new axiom, however, are considerably simpler and easier to discover, and make it possible to condense into one proof many different proofs.

However, it's at least prima facie appealing to expect central principles of set theory which are used without comment to have intrinsic justification, and this expectation seems common in other areas of mathematics. For example, it seems that everything we want to say about the natural numbers (in the language of arithmetic) follows from (say) our second-order conception of the natural numbers.

If it turns out that adequate intrinsic justification cannot be given, it might be reasonable to accept extrinsic justification (for we do this in the sciences, after all). And perhaps we will reach a point with, e.g., large cardinal axioms, where extrinsic justification is all we can provide. However, one might hope to do better with regard to the ZF axioms, which are treated as quite secure and used to provide a foundation/explication of normal mathematical claims that we are very confident in. Even if appeal to the fruitful good consequences of Replacement provides some justification for believing it, this doesn't secure the kind of intrinsic convincingness we usually expect (and hope for) from mathematical axioms.

Second, Potter (2004) suggested justifying Replacement by appeal to a kind of inference to the best explanation along the following lines. Russell's paradox tells us

---

[12] Potter supports this assessment by quoting: "Skolem . . . gives as his reason that 'Zermelo's axiom system is not sufficient to provide a complete foundation for the usual theory of sets', because the set $\{\omega, P(\omega), P(P(\omega)), \ldots\}$ cannot be proved to exist in that system; yet this is a good argument only if we have independent reason to think that this set does exist according to 'the usual theory', and Skolem gives no such reason. Von Neumann's . . . justification for accepting Replacement is only that, 'in view of the confusion surrounding the notion 'not too big' as it is ordinarily used, on the one hand, and the extraordinary power of this axiom on the other, I believe that I was not too crassly arbitrary in introducing it, especially since it enlarges rather than restricts the domain of set theory and nevertheless can hardly become a source of antinomies'."

that not all pluralities of objects can form a set (there isn't a set of all sets that aren't members of themselves). So, if there are any sets, there should be a principled division between those pluralities of objects which can form sets and those which can't. But sets don't have that many features. So (one might think) size is the only natural choice for the limitation on what pluralities count as sets and it should be the *only* such limitation (Potter 2004). As Michael Potter puts it, we should accept the Size Principle: "If there are just as many Fs as Gs, then the Fs form a collection if and only if the Gs do" (which implies Replacement), because:

[A] collection is barely composed of its members: no further structure is imposed on them than they have already. So ... what else could there be to determine whether some objects form a collection than how many there are of them? What else could even be relevant? (Potter 2004: 231)

I don't find this inference to the best explanation very convincing because sets do have *some* other features than their size which could be used to explain why certain pluralities of sets fail to form a set in a style analogous to Potter's explanation for this fact.

In particular, note that on the iterative hierarchy conception of sets (which Potter accepts) each set will have the property of first being generated at some ordinal level α. This feature of sets is a fairly natural and principled one. One can think of it as reflecting how many layers of indirect and metaphysically derivative object existence (given the common idea that sets are in some sense metaphysically dependent on their elements, not vice versa)[13] one has to go through to arrive at that set.

So, rather than hypothesizing (with Potter) that the iterative hierarchy of sets stops at a certain point because ascending any further would require collecting objects which are *too plentiful* to form a set, couldn't we just as well hypothesize that the iterative hierarchy of sets stops somewhere because any further sets formed would have to occur *too high up* in an iterative hierarchy (i.e., one would have to ascend through too many layers of abstraction/metaphysical dependence to form a set from the relevant elements)? To the same (rather fanciful) extent that we can imagine that the rubber band holding together the elements of a sets just happens to be too small to collect any plurality of elements of a certain size κ, we could imagine that the power of lower-level sets to ground the existence of higher-level sets and thereby indirectly to ground the existence of still higher-level sets etc. eventually becomes too attenuated to allow any further sets to be formed at some height α.

So, if we're just accepting Replacement on the basis of inference to the best explanation, how do we know there's an upper bound to the *sizes* sets can have vs. an upper bound to the *rank* they can have? One might also object to Potter's methodology more generally, on the grounds that even philosophers who are happy to use the

---

[13] See, for example, Bliss and Trogdon (2016) for the development of the intuition that the existence of Socrates's singleton is to be grounded in the existence of Socrates and depends on that, in a way that the existence of Socrates does not depend on the existence of his singleton, and use of this intuition to motivation a notion of grounding which is distinct from metaphysically necessary covariation and supervenience.

kind of metaphysical inference to the best explanation suggested by Potter's justification don't usually take applying this method to justify the great confidence and certainty we feel in typical mathematical results.

Third, one can provide a kind of justification for Replacement by noting it follows from a set-theoretic reflection principle.[14] I take this proposal (and the one that follows) to typically arise from the attempt to find a unified conception of the sets from which the ZFC axioms follow (whether or not that conception is obviously true or coherent) rather than any attempt to derive the axiom of Replacement from something that seems more obviously true. But I will discuss both proposals for completeness.

Informally, the idea behind reflection principles is that the height of the universe is "absolutely infinite" and hence cannot be "characterized from below." A specific reflection principle will assert that any statement ϕ in some language that's true in the full hierarchy of sets V is also true in some proper initial segment V_α. This ensures that one cannot define V as the unique collection which satisfies ϕ (or the shortest such collection) since there will be a proper initial segment V_α of V that satisfies ϕ.

More formally, once accepts first-order reflection/second-order reflection etc. insofar as one accepts all instances of the following schema, where ϕ is a first-order/second-order etc. formula:

Reflection Schema For any objects $a_1, \ldots, a_n$ in V_α, we have
$\phi(a_1, \ldots, a_n) \leftrightarrow V_\alpha \models \phi(a_1, \ldots, a_n)$.

If one accepts first-order reflection, then one can justify Replacement.[15]

This third strategy (justification by appeal to a reflection principle) is *somewhat* attractive. For, as Koellner (2009) reviews, one can motivate reflection principles[16] by Gödel's idea that the total hierarchy of sets (V) should be impossible to define. For reflection principles (in effect) say that anything that's true of the whole hierarchy of sets will also be true in some proper initial segment of it. If some instance of a reflection principle failed (so there was some fact about the whole hierarchy of sets that didn't reflect down to be true of a proper initial segments of the sets), then we could (in a sense) define the hierarchy of sets by saying it is the shortest[17] iterative hierarchy structure satisfying this claim. Gödel writes:

Generally, I believe that, in the last analysis, every axiom of infinity should be derivable from the (extremely plausible) principle that V is indefinable, where definability is to be taken in [a] more and more generalized and idealized sense.[18]

I admit that the idea in the quote above has a kind of elegance and provides a kind of internal justification for reflection (as opposed to the external justification by consequences evoked above).

---

[14] My summary of this approach follows Koellner (2009).   [15] See, for example, Button (n.d.).

[16] Different reflection principles correspond to different classes of sentences being reflected. For instance, you might think only first-order sentences reflect or first-order formulas with parameters or second-order sentences etc.

[17] That is, the sets satisfy the non-reflected claim but no initial segment does.

[18] This is quoted from Wang (1998) in Koellner (2009).

However, it's not obvious (or not as obvious as we'd naively hope foundational axioms for mathematics could be) that there could be a structure satisfying the intuition behind reflection (or even second-order reflection) together with our other expectations about the hierarchy of sets (e.g., the other ZFC axioms, IHW).

Also, to the extent that Gödel's idea in the quote above motivates the first-order Reflection principle used to justify Replacement above, it would seem to also motivate third-order reflection, some instances of which (as Koellner notes in the article cited above) have been shown to be inconsistent (Reinhardt 1974). So, one might think that justifying Replacement by merely noting that it follows from Reflection doesn't provide enough justification.

Fourth, philosophers like Boolos (1971a, 1989) justify Replacement from a size principle. (Speaking informally), the idea is to say that some plurality of objects forms a set if and only if it is "small," where the latter means that its members can't be bijected with the total universe. This principle justifies Replacement, because the set you get by applying Replacement to a set u must be the same size as u or smaller.

But, just as with Reflection, it's not as clear as one would like that it would be coherent for there to be a structure with the intended width of the hierarchy of sets that satisfies this property together with the axiom of infinity.

A fifth style of justification considered by Button (n.d.) derives Replacement from the following principle:

Stages-are-super-cofinal. If A is a set and $\tau(x)$ is a stage for every $x \in A$, then there is a stage which comes after each $\tau(x)$ for $x \in A$.

Button notes that we can motivate the following formal claim by appealing to the informal principle below, which he says is "consonant with" the cumulative-iterative conception of set:

Stages-are-inexhaustible. There are absolutely infinitely many stages; the hierarchy is as tall as it could possibly be.

However, I don't currently grasp the kind of modality that's intended to be evoked by the term "possibly" in *stages-are-inexhaustible*. Earlier in this chapter I've tried to invoke an intuitive sense of possibility on which there *couldn't* be an iterative hierarchy "as tall as it could possibly be" (for any structure of objects satisfying IHW, there could be a strictly taller one). And we will see below that Potentialists like Putnam, Parsons, Hellman, Linnebo and Studd have appealed to notions of logical or interpretational possibility which (they think) conform to this intuition.

And without the additional justification provided by the informal principle *stages-are-inexhaustible*, we find ourselves in an epistemic situation similar to just taking Replacement or some form of Reflection as an axiom, as regards *stages-are-super-cofinal*. It's not implausible, but also not seemingly obvious/clearly true that it would be logically coherent for there to be an iterative hierarchy that satisfies the relevant closure principle. Thus, I don't think merely pointing out that *stages-are-super-cofinal* implies Replacement doesn't suffice to justify the latter from principles that seem clearly true.

So, to summarize the discussion of different Actualist strategies for justifying Replacement above, we get the following picture. In order to justify the level of confidence we have in set theory, and particularly Replacement, as well as for aesthetic reasons, we would like our set-theoretic axioms to follow from some simple, intuitive conception which strikes us as prima facie clearly logically coherent.

For instance, we think of number theory as describing the sequence built by starting at 0 and continuing to add successors "as long as is needed to ensure that there is no last natural number, but no longer" in a sense which can be cashed out via the second-order axiom of induction. And we can think of the real numbers as describing a line extending to infinity in both directions without gaps (i.e., such that it's impossible to add any further "number" anywhere on the line without it being equal to a real[19]). In both these cases, we seem to have a unified, precise and intuitively consistent conception of the relevant mathematical structure, from which our first-order axioms describing the natural numbers/real numbers flow.

The iterative hierarchy idea sketched in Section 2.1 plausibly specifies the width of the hierarchy of sets in a way that's logically coherent (on its own). But just assuming that the sets satisfy this width requirement (or even that adding that there's no last stage to the hierarchy of sets) doesn't suffice to justify Replacement. Adding principles like Reflection or Boolos' size principle to our conception would ensure that our conception of the intended structure of the sets implies Replacement (and hence perhaps that if there are sets, then they satisfy Replacement). However, we have little or no reason to think this enlarged conception is coherent. So, it provides little justification for thinking that the axiom of Replacement is even consistent with the other principles about the hierarchy of sets (hence little justification for thinking it's true).

In the next few chapters, I will argue that adopting a Potentialist approach to set theory lets us do better with regard to both the arbitrariness and justification problems above.

2.5   Indefinite Extensibility

But, before I go on to the development and defense of Potentialism, let me end by quickly saying something about the limits of the argument discussed in this chapter.

Many other philosophers interested in Potentialism about the height of the hierarchy of sets, such as I will develop in response to the arbitrariness worry, have also explored more general versions of Potentialism, which go further and reject the idea that we have a definite conception of the structure of the natural numbers or the width of the hierarchy of sets. Thus, one might wonder if there is a principled reason for taking a Potentialist approach to the height of the hierarchy of sets but not to the width of the hierarchy of sets or the natural numbers.

---

[19] One can think of a Dedekind cut which doesn't correspond to a real number as a kind of gap, i.e., a vertical line passing through the x-axis that somehow misses every real number.

In the remainder of this chapter, I will answer this question by clarifying why I think the motivation for height Potentialism about set theory doesn't generalize in the ways just mentioned. I will contrast the claims I've made about our *lacking a coherent categorical* conception of an Actualist hierarchy of sets above with Dummett's famous – and famously obscure – remarks about indefinite extensibility.

2.5.1 Height Potentialism and No More

While one can certainly doubt that we can uniquely refer to the intended structure of the natural numbers or the subsets of a given set there is no (similarly compelling) paradox like Burali-Forte that arises from assuming that we can.

Here's another way of thinking about the disanalogy. One can fairly concretely imagine an ordinal-like-object above any well-ordered plurality of ordinals and a layer of set-like-objects above any plurality of sets satisfying IHW. We can specify exactly how $\le$ and $\in$ would relate the new sets/ordinals to all the old sets/ordinals previously considered so as to form a new structure satisfying IHW equally well. And the structure we imagine forming by extending any given plurality of ordinals has as good a claim to contain all the objects that satisfy our conception of "the ordinals"/"the sets" as the original structure, if our conception after rejecting the naive height principle is just IHW. And in any case our conception of the ordinals/sets doesn't seem to include any (coherent) negative conditions, which say that the height of the hierarchy must stop at a certain point.

But we can't do the same thing with our concepts of "full" second-order quantification (aka arbitrary subsets of a given collection), natural number and real number. Perhaps, in a sense, it's intuitive that, for any collection of natural numbers (finite or infinite), we can imagine a strictly larger *vaguely* number-like object. For we can always imagine adding (something like) a successor or a limit ordinal after all numbers within any collection of numbers. However, our grasp of the natural numbers does very centrally include such a principle saying the numbers must stop at a certain point, namely the second-order induction axiom! We think the numbers are (so to speak) as *few as can be*[20] while containing 0 and the successor of everything they include and that for this reason, any property which applies to 0 and applies to the successor of everything it applies to must apply to all the numbers. The same goes for the concept of full second-order quantification/all possible subsets of a given collection. We have no positive intuition about how to generate, for any given collection of sets of cats, a new set-of-cats-like object which is distinct from all the ones previously considered.[21]

---

[20] Here I mean "few" in an order type sense, not a cardinal sense. Maybe it would be better to say that the natural number structure is as short/small as can be while satisfying this condition.

[21] Perhaps Hamkins' radical multiverse proposal provides a way of developing the latter counter-intuitive idea. But see my discussion of Hamkins in Section 9.4.

2.5.2   Contrast with Dummett

It may be helpful at this point to contrast my arbitrariness problem for Actualism with Michael Dummett's influential arguments about indefinite extensibility. In "What Is Mathematics About?," Dummett (1993) raises something very much like the Burali-Forti worry I pressed above concerning the height of the hierarchy of sets:

If it was ... all right to ask, "How many numbers are there?", in the sense in which "number" meant "finite cardinal," how can it be wrong to ask the same question when "number" means "finite or transfinite cardinal?" A mere prohibition leaves the matter a mystery. It gives no help to say that there are some totalities so large that no number can be assigned to them. We can gain some grasp on the idea of a totality too big to be counted, even at the stage when we think that, if it cannot be counted, it does not have a number; but, once we have accepted that totalities too big to be counted may yet have numbers, the idea of one too big even to have a number conveys nothing at all. And merely to say, "If you persist in talking about the number of all cardinal numbers, you will run into contradiction," is to wield the big stick, not to offer an explanation.[22]

And one might say that both of us reject standard Actualist set theory on the grounds that our conception of sets is, in some sense, "indefinitely extensible." However, Dummett is concerned with indefinite extensibility in a different sense than I am. Specifically, I reject standard (Actualist) Platonism about set theory because our concept of sets and ordinals is "indefinitely extensibile" in the following strong sense (if we take the natural conception of set that remains, once we reject the naive and paradoxical conception that the sets go "all the way up," to be IHW):

Strong Indefinite Extensibility We have a positive intuition that for any hierarchy of sets/ ordinals *there could be*, there could be a strictly larger one which matches our conception of the sets (IHW)/ordinals equally well.

In contrast, Dummett seems to reject standard Platonist set theory because our concept of sets is "indefinite extensible" in this weaker sense:

Weak Indefinite Extensibility For any collection of numbers/sets/ordinals *we can form a definite conception of* (which Dummett says he will start by presuming means any *finite* collection!) this collection can be extended so as to contain extra things which would also fall under our conception of that structure

Dummett writes, "[A]n indefinitely extensible concept is one such that, **if we can form a definite conception of** a totality all of whose members fall under the concept, we can, by reference to that totality, characterize a larger totality all of whose members fall under it" (Dummett 1993: 440; emphasis added).

To support this reading, consider how Dummett argues that the concepts of natural numbers are "indefinitely extensible" by (seemingly) assuming that all totalities of numbers we can form a definite conception of collect numbers from 0 to n for some n. His story

---

[22] Dummett (1993: 439).

about how to extend an arbitrary totality of natural numbers (that we can definitely conceive of) is simply the following:

given any initial segment of the natural numbers, **from 0 to n**, the number of terms of that segment is again a natural number, but one larger than any term of the segment.

Similarly, the argument Dummett takes to show that our concept "real number" is indefinitely extensible is simply Cantor's diagonal argument that any countable plurality of real numbers must be leaving some real numbers out.

Indeed Dummett explicitly notes that he's making these assumptions (of finiteness and countability) in the quote below and (unsurprisingly) recognizes they will strike opponents as question-begging:

A natural response is to claim that the question has been begged. In classing *real number* as an indefinitely extensible concept, we have assumed that any totality of which we can have a definite conception is at most denumerable; in classing *natural number* as one, we have assumed that such a totality will be finite. Burden-of-proof controversies are always difficult to resolve, but, in this instance, it is surely clear that it is the other side that has begged the question. (Dummett 1993: 443)

Dummett goes on to defend this burden of proof claim by arguing that it's mysterious how a definite conception of an infinite structure could be communicated, and the burden of showing such communication is possible falls on his opponent.

I won't try to adjudicate this dispute here. Much can and has been said about whether this succeeds and how to understand Dummett's infamously "dark" (Rumfitt 2015) notion of indefinite extensibility (Dummett 1991).

Instead, I merely want to note that Dummett's arguments for the (weak) indefinite extensibility of the natural numbers and real numbers don't even pretend to show the strong indefinite extensibility of these notions. They don't pretend to show that, for *any* totality of objects related by some relation R in the way we believe the natural numbers to be related by successor, it would be it would be intuitively possible/logically coherent to have a strictly larger structure that accords with our conception of the natural numbers equally well. Thus, Dummett's reason for worrying about the sets arguably applies to the natural numbers and real numbers (any *finite* collection of these will be missing a number which could be added) etc. while (we've just seen above that) mine doesn't.

Philosophically speaking, I suspect these different "indefinite extensibility" worries arise from different philosophical projects and background assumptions as follows.

I take both the naive intuition that we mean something definite by both "all possible subsets" and "all the way up" at face value until Burali-Forti paradox shows the latter is contradictory. Since no analogous paradox seems to arise for "all possible subsets," I'm happy to invoke this notion in expressing a conception of the natural numbers, etc.

In contrast, Dummett starts from a more skeptical/cautious position and asks to be shown how one could "convey" a definite concept of structures to someone who starts out only understanding finite collections. And he prima facie doubts that you could do so by, e.g., giving an operation like adding one and talking about closing under it or relating your natural number concept to reference magnetic notions of second-order quantification or logical possibility.[23]

Thus, I think, the fact that Dummett's more skeptical worry applies more widely than the Burali-Forti driven worry I've pressed is unsurprising.

---

[23] Perhaps we can latch onto a notion of logical possibility which (we will see below) suffices to categorically describe the numbers and sets in the same way (whatever it is) that we can latch on to a notion of objective physical possibility/law. For example, it might be that we get both notions by making certain core good inferences (e.g., the actual to possible, Axiom 8.1, and uniform relabeling, Axiom 8.5, principles, I introduce below in the case of logical possibility, and some other kind of extrapolation in the case of physical possibility) which in a way under-determine which modal notion we mean and then benefiting from reference magnetism. Thus, I suspect that Dummett's worry either (despite protests to the contrary) comes down to an argument from some principle of manifestability which would call reference to realist physical possibility/law facts into doubt, or reduces to my worry about the height of the hierarchy of sets. However, I won't pursue this argument here because my present aim is only to explain how my worry differed from Dummett's, not to answer his worry.

3 Putnamian Potentialism: Putnam and Hellman

Let us now turn to Potentialism, a different approach to set theory. There are two broad schools of Potentialism which, following Linnebo (2018b), I will call Putnamian and Parsonian Potentialism, and discuss in this book.

In this chapter I will present Potentialism from a Putnamian point of view. Later, in Chapter 5, I will discuss rival Parsonian proposals. Unlike Actualists, Putnamian Potentialists don't take set theory to describe actual or possible existence of special objects called "sets."

In a nutshell, Potentialists interpret mathematicians who appear to be quantifying over the sets as really talking about the possibility and extensibility of structures satisfying the iterative hierarchy conception of sets discussed in Chapter 2. We might say that Potentialist translations talk about the possibility of there being (objects with the structure of) standard width initial segments V_α of the total hierarchy of sets V, and how some such initial segments could be extended by longer initial segments. They don't interpret set theorists as quantifying over any collection of existing objects, or even as talking about what follows from some axioms describing the supposed structure of the sets. Instead, they systematically interpret mathematical utterances which appear to quantify over the sets as having a much more complicated logical form.

Crudely speaking, the Potentialist will interpret singly quantified existential claims $(\exists x)(\phi(x))$ in set theory (e.g., $(\exists x)(x = x)$), as saying (something like) that it's possible for there to be a standard width initial segment of the hierarchy of sets containing an object x satisfying ϕ (in this case $x = x$). And they will interpret set-theoretic claims of the form $\forall x \phi(x)$, where ϕ is quantifier free, (e.g., $(\forall x)(\neg x \in x)$), as saying (something like) that it's necessary that any object x within a standard width initial segment of the hierarchy of sets has the property ϕ.

What about set-theoretic claims involving nested quantification? The Potentialist will interpret statements of the form $(\forall x)(\exists y)\phi(x, y)$ (where ϕ has no quantifiers) as saying (something like): it's necessary that for any standard width initial segment V and object x within it, it's possible to have larger initial segment V' extending V, containing an object y, such that $\phi(x, y)$. The same pattern continues for more logically complex sentences.

In this chapter I'll discuss how Putnam and Hellman have developed Putnamian Potentialism so far and how it promises to answer some of the problems for traditional Actualist set theory discussed in Chapter 2. I'll then raise, and begin to answer, some worries for these versions of Potentialism, about how to spell out the relevant notions of possibility and extensibility.

3.1   Putnam

Hillary Putnam (1967) sketches a way of thinking about set theory in terms of modal logic: as talk about what "models" of set theory are, in some sense, *possible* and how such models can be extended.

He introduces a notion of being a standard model of set theory, which is a model of set theory closed under subsets, i.e., a hierarchy of sets having full width and no infinite descending chains under $\in$.[1] Putnam says that we can "make this notion concrete" by thinking of models as physical graphs consisting of pencil points (or the analog of pencil points in space of some higher cardinality) with arrows connecting these pencil points. He "ask[s] the reader to accept it on faith" that we can express the claim that some model is standard in this way "using no 'non-nominalistic' notions except the '◊'" (where ◊ denotes a notion of possibility to be discussed below).

With this notion of a concrete model in place, Putnam suggests that we can understand set-theoretic statements as claims about what such models are possible, and how they can be expanded. For example, he proposes that we can paraphrase a set-theoretic statement of the form "$(\forall x)(\exists y)(\forall z)\phi(x, y, z)$," where ϕ is quantifier free, as saying that, if G is a standard concrete model, and p is a point within G, then it is possible that there is a model G' which extends G, and a point y within G' such that necessarily, for any model G'' which extends G' and contains a point z, $\phi(x, y, z)$ holds within the concrete model G''. And we can treat arbitrary quantified statements in set theory in an analogous fashion.

Putnam then suggests that adopting this Potentialist approach to set theory can help us dispel the kind of arbitrariness and indefinite extensibility worries I discussed in Section 2.2. The Potentialist can understand set-theoretic talk without imposing or positing arbitrary limits on the size of structures (as we would do if we just stipulated a point at which the hierarchy of sets stopped or inferred that it must stop somewhere) in a way that seems faithful to our intuitions about the generality of set-theoretic reasoning.[2] As Putnam puts it:

---

[1] Specifically, Putnam writes "[A concrete] model will be called standard if (1) there are no infinite-descending 'arrow' paths; and (2) it is not possible to extend the model by adding more 'sets' without adding to the number of 'ranks' in the model. (A 'rank' consists of all the sets of a given-possibly transfinite-type. 'Ranks' are cumulative types; i.e., every set of a given rank is also a set of every higher rank. It is a theorem of set theory that every set belongs to some rank.)"

[2] In particular (before thinking about the paradoxes), we'd hoped for set theory to be general in the sense that every possible structure will have a copy somewhere in the sets.

[W]e have a strong intuitive conviction that whenever As are possible, so is a structure that we might call "the family of all sets of As." ... from the standpoint of the modal-logic picture ... the Russell paradox ... shows that no concrete structure can be a standard model for the naive conception of the totality of all sets; for any concrete structure has a possible extension that contains more "sets." (If we identify sets with the points that represent them in the various possible concrete structures, we might say: it is not possible for all possible sets to exist in any one world!) Yet set theory does not become impossible. Rather, set theory becomes the study of what must hold in, e.g., any standard model for Zermelo set theory. (Putnam (1967: 311)

I think Putnam is right that his proposal indicates an appealing style of response to the worries about arbitrary stopping points for the hierarchy of sets indicated above. And (as we will see) it has inspired many other philosophers. However, this proposal is (explicitly) sketchy on certain formal and philosophical points. For instance, Putnam doesn't provide any criteria for what it would take for some physical graph (formed of pencil points and the like) to form "a standard model for Zermelo set theory" but rather (as we saw above) asks the reader to "accept it on faith that the statement that a certain graph *G* is a standard model for Zermelo set theory can be expressed using no 'non-nominalistic' notions except the '◊'."[3]

And, philosophically, Putnam says rather little about the modal notion he intends to capture with the ◊. Indeed, he seems to vacillate between a purely mathematical understanding of necessity and a physical understanding.

At some points he seems to have a notion of explicitly "mathematical possibility" in mind. For example, he writes (brackets in original): "assuming that the notions of mathematical possibility and necessity are clear [and there is no paradox associated with the notion of necessity as long as we take the '◊' as a statement connective (in the degenerate sense of 'unary connective') and not ... as a predicate of sentences], I wish to employ these notions to try to give a clear sense to talk about 'all sets'" (Putnam 1983a). However, at earlier points in the same article Putnam seems to appeal to something more like metaphysical possibility or a priori conceivability. For he writes as if constraints on how we can conceive the structure of physical space might block the possibility of relevant models, and makes assumptions about this which philosophers such as Parsons (2007) and Tait (2005) have been unwilling to grant, e.g., Putnam says: "I assume that there is nothing inconceivable about the idea of a physical space of arbitrarily high cardinality; so models of this kind need not necessarily be denumerable, and may even be standard."

Additionally, Putnam advocates Potentialism as merely one possible and helpful "perspective" on mathematics, and claims that it is, in some sense, equivalent to a more familiar Actualist understanding of set theory, which only appears to be incompatible with it. But cashing this idea out clearly requires serious and disputable metaphysics.[4]

Furthermore, it's not clear that saying both perspectives are equally good is compatible with honoring Putnam's Potentialism-motivating intuition that "whenever As are possible, so is a structure that we might call 'the family of all sets of As'" (Putnam 1983b). We seem forced to *either* say that the idea that for any structure there could be a

---

[3] Here nominalistic notions are ones that aren't committed to the literal existence of mathematical objects.
[4] See, for example, John Burgess' vigorous objections to Putnam's stance in Burgess (2018).

larger one is only true "from the Potentialist perspective" on mathematics or to say that it is true simpliciter, even from the Actualist perspective.

The former position can feel a little mysterious and unsatisfying, but the latter is uncomfortable for two reasons. First (like more straightforward forms of Actualism), it involves positing arbitrariness in mathematical reality by saying the Actualist hierarchy of sets just happens to stop somewhere, though it could go on further. Second, it's not clear (even at a very loose intuitive level) how talking about any such Actualist hierarchy could be equivalent to a practice of modal set theory which considers arbitrary logically possible extendibility.[5]

3.2 Hellman

Geoffrey Hellman (1994; 1996; 2011) develops Putnam's ideas about Potentialist set theory as part of a larger purely Nominalist philosophy of mathematics, in a way that addresses or avoids some of the worries above.

First, Hellman drops Putnam's suggestion that Actualist and Potentialist approaches to set theory are (somehow) two equally good perspectives on the same thing. Instead, he simply advocates and develops a Potentialist understanding of set theory. I will follow suit.

Second, Hellman provides a somewhat clearer picture of what the key modal notion $\Diamond$ in (his version of) Putnam's Potentialist set theory is supposed to mean, saying that it's supposed to express a primitive modal notion of logical possibility. However, he does relatively little to describe this notion. He does say that, "[when evaluating logical possibility] we are not automatically constrained to hold material or natural laws fixed." So, it may be logically possible that $(\exists x)(\text{pig}(x) \land \text{flies}(x))$, but physically impossible. And he adds that, "we are free to entertain the possibility of additional objects – even material objects – of a given type." So, for example it's logically possible that there are infinitely many objects even if there are actually only finitely many objects. And it's logically possible for there to be say $2^{2^{\omega}}$ cats, even if it's not metaphysically possible for there to be so many cats. This (arguably) lets us avoid concerns about limitations on the cardinality of space unduly limiting the range of possible models considered above. Beyond this remark, however, Hellman just suggests that his applications of logical possibility will make the notion he has in mind clear.

Hellman also does a lot to fill in the other promissory notes left by Putnam's sketch. He cashes out Putnam's appeal to "standard models" of set theory by saying that standard models are models which satisfy ZFC_2 (i.e., the version of standard ZFC set theory which replaces the inference schemas of Replacement and comprehension with corresponding second-order axioms).[6]

---

[5] Perhaps one could say that the Actualist hierarchy is the smallest standard width structure whose truth conditions for all first-order logical claims agree with those provided by the Potentialist set theory.

[6] So, for example, ZFC expresses comprehension via an axiom schema which contains an axiom for every formula $\Diamond$ in the language of set theory. In contrast, by using second-order logic one can state a single

Two details of how Hellman spells this idea out will be particularly important to note, insofar as I think they can raise problems and my own Potentialist paraphrases will follow Putnam rather than Hellman in these matters.

3.2.1 Which Hierarchies?

First, Where Putnam spoke of models of "Zermelo set theory" (which doesn't include Replacement), Hellman talks about models satisfying second-order ZFC, and hence Replacement. That is, Hellman takes the initial segments whose possible extensions Potentialism considers to *themselves* satisfy ZFC_2.

Now if one accepts the relevant large cardinal axioms, then there's a sense in which this change makes no difference. For (it turns out) that Potentialist translations taking initial segments to satisfy ZFC_2 will be logically equivalent to translations involving initial segments that satisfy much weaker requirements like IHW_2 or ZFC. But it may make a difference to our current project of justifying the Potentialist version of ZFC from intuitively compelling principles.[7] To infer even the simplest existential claim in set theory (e.g., to say that there is a set that is self-identical), Hellman would need to know that it was logically possible for a structure to satisfy ZFC_2. And the logical coherence of a hierarchy of sets satisfying second-order ZFC is by no means obvious, especially in the context of our current doubts about Replacement.

One might also feel that requiring the initial segments being extended to satisfy ZFC_2 or even constitute a "standard model of Zermelo set theory" (rather than merely satisfying our conception of being an intended width hierarchy IHW above, e.g., IHW_2) is slightly unnatural. In Chapter 2, I tried to paint the following picture. We seem to have a precise and consistent conception of the intended width of the hierarchy of sets, but (as we see when deriving contradiction from the Naive conception of absolute infinity in Section 2.2) no such conception of its intended height. Now one might say: the point of Potentialism as a solution to the arbitrariness problem, is to solve this problem of heights. So Potentialist set theory should talk about how iterative hierarchies of standard width could be extended, rather than imposing any height constraints. But I admit that perhaps this is a matter of taste.

3.2.2 Appeal to Second-order Logic or Plurals

Hellman (1994) also makes a second change that, which I want to highlight because my own proposal will wind up being closer to Putnam's original proposal in this regard. When Putnam talks about the modal perspective on mathematics, he considers possibility of objects being related by **specific first-order relations** as per certain set-theoretic axioms. So, for example, we might consider the possibility that the pencil

---

comprehension axiom: $(\forall x)(\forall C)(\exists y)(\forall z)(z \in y \leftrightarrow z \in x \wedge C(z))$. The same goes for the first-order axiom schema of Replacement and its second-order analog.

[7] I gather Hellman (2020) chooses to go this way in an attempt to bring out a kind of analogy between Replacement and large cardinal axioms, something which I don't attempt here.

points form an intended model of Zermelo set theory when considered under the relation "an arrow points from ... to ...." If we followed Hellman in requiring our hierarchies to satisfy ZFC_2, this would amount to saying that all axioms of ZFC_2 become true when you replace "set" with "point" and "element of" with "an arrow points from ... to ...." However, he notes that any relations of the right arity will do.[8] We could translate a given sentence of set theory equally well by talking about how it would be logically (or logico-mathematically in whatever sense Putnam has in mind) possible for the pencil points to arrow one another or the angels to admire one another, the dogs to sniff each other, or any other kind of object under some two-place relation.

In contrast, Hellman (1994) interprets set-theoretic claims purely in terms of second-order quantification. That is, instead of saying something about how it's logically possible for penciled points to arrow one another, we talk about the possibility of there being second-order class and relation objects X and f such that $ZFC_2[set/X, \in/f]$ (where this means that ZFC_2 holds if you replace the predicate "set" with membership in X and the predicate $\in$ with membership in the two place relation f^9).

In a later work, Hellman (1996) modified this view slightly, as motivated by his Nominalism and famous Quinean sentiment that second-order logic is ontologically committal (Quine 1970) (so accepting second-order comprehension commits one to abstract objects). He notes that quantifying over all pluralities xx automatically lets you simulate second-order X quantification, and that (if you can make certain assumptions about the size of the universe and use mereology) you can also simulate second-order relation or function quantification f via the strategy indicated in the appendix of David Lewis' *Parts of Classes* (Lewis 1991). Accordingly, he proposes to rewrite his paraphrases in terms of plural quantification and a notion of parenthood, rather than second-order quantification.[10]

---

[8] For example, Putnam (1967: 10–11) writes: "Let 'AX' abbreviate the conjunction of the axioms of the finitely axiomatizable subtheory of first-order arithmetic just alluded to. Then Fermat's last theorem is false just in case '$AX \supset \neg$ Fermat' is valid, i.e., just in case

$$(1)\ \Box\ (AX \supset \neg\ \text{Fermat})$$

Since the truth of (1), in case (1) *is* true, does not depend upon the meaning of the arithmetical primitives, let us suppose these to be replaced by 'dummy letters' (predicate letters). To fix our ideas, imagine that the primitives in terms of which AX and $\neg$Fermat are written are the two three-term relations 'x is the sum of y and z' and 'x is the product of y and z' (exponentiation is known to be first-order-definable from these, and so, of course, are zero and successor). Let $AX(S, T)$ and $\neg$Fermat(S, T) be like AX and $\neg$Fermat except for containing the 'dummy' triadic predicate letters S, T, where AX and $\neg$Fermat contain the constant predicates 'x is the sum of y and z' and 'x is the product of y and z.' Then (1) is essentially a truth of pure modal logic (if it is true), since the constant predicates occur 'inessentially'; and this can be brought out by replacing (1) by the abstract schema: (2) $\Box[AX(S, T) \supset \neg\text{FERMAT}(S, T)]$ – and this is a schema of pure first-order modal logic."

[9] While f is traditionally used to denote a function here, Hellman allows f to be any two-place relation.

[10] In even later work, Hellman (2011) switches to a two-sorted view, where we have ordinals o understood in something akin to the Parsonian sense discussed in Chapter 5, and plural quantification over both pluralities of ordinals oo and pluralities of objects xx which will play the role of sets in satisfying ZFC_2 and forming a hierarchy of sets with layers corresponding to some ordinals oo.

I have some doubts about the success of this move. In particular, I'm not convinced that this use of mereology to simulate second-order quantification can be combined with taking the ◊ to express logical possibility rather than metaphysical possibility. For reasons that will become clear below,[11] I don't think that the axioms of mereology are logically necessary. If logical possibility ignores metaphysically necessary constraints on how many concrete objects can exist in space and time, shouldn't it ignore the metaphysically necessary laws of mereology too? Thus, I think that employing this strategy to formulate Potentialist set theory (rather than just modally paraphrasing talk of smaller structures like the numbers and the reals as Hellman (1994) suggests) would reawaken the problems about the metaphysical possibility of arbitrarily large cardinalities of objects noted above. And, as we will see, my approach also eliminates the use of second-order quantification in favor of a notion of logical possibility with, arguably, a stronger claim to ontological innocence than plural quantification.

But I won't dwell on this issue more here,[12] as I'm not a Nominalist myself.

3.3   Remaining Problems

Altogether, I think adopting Potentialist set theory in the manner developed by Putnam and Hellman has significant appeal. As noted above, it helps solve the arbitrariness problem which Actualists face regarding the height of the set-theoretic hierarchy. But with this picture of the current state of Putnamian Potentialist set theory in mind, I'll now note three lingering problems about Hellman's system.

First, as mentioned above, there's a concern about how precisely to understand Putnam's notion of possibility and Hellman's notion of logical possibility. I'll argue in Section 4.1.3 that independent work in the philosophy of logic provides significant clarification of the relevant notion of logical possibility, and an attractive way of spelling out the relevant notion.

3.3.1   Quantified Modal Logic

Second, there's a problem about the infamous controversialness of quantified modal logic. Putnam and Hellman both "quantify in" to the ◊ of logical possibility (or whatever other modality is used to cash out Potentialist set theory). That is, they use sentences like $\exists x \Diamond R(x)$, where the logical possibility operator is applied to a formula with free variables. But there are significant controversies about the truth-value (and/or meaning) of even very simple sentences involving quantifying-in to the ◊ of logical or metaphysical possibility. Additionally, as I'll suggest below, working in a language that allows both quantifying-in to the modal operator and standard FOL inferences can

---

[11] If we take logical possibility to be interdefinable with validity in the way advocated in Section 3.1, the logical contingency of mereology seems to follow.
[12] I do consider switching to my Potentialist framework as a friendly suggestion to advocates of Nominalism (as I suggest in Berry 2018a).

make the consequences of axioms difficult to survey, and thus makes it harder to see that these axioms are clearly true (as needed for our foundational project).

These roadblocks to finding uncontroversial modal principles that can easily be seen to be true for the language of Hellman's Potentialist paraphrases make using these paraphrases inconvenient for the key project of this book: justifying set theory via principles that are as intuitively obvious seeming as possible.

I'll use this worry about the meaning and truth-value of basic claims in the language of quantified modal logic, to motivate a switch to my preferred version of Putnamian Potentialism in Chapter 4.

3.3.1.1   Quinean Qualms

Most radically, Quine famously argued against quantifying-in to modal contexts all together. I take Quine's main problem with quantifying-in, in Quine (1953), to be that he dislikes the "Aristotelian essentialism" of saying that some properties belong to an object like the number 7 essentially (e.g., being less than 9) while others apply only contingently (e.g., being the number of planets). After all, taking there to be such an abundance of facts about essences can seem like positing a bunch of arbitrary and unneeded metaphysical facts. But perhaps these concerns are less severe if we specify that we're only talking about logical possibility, because objects' logical essences will be (somehow) "minimal."

3.3.1.2   Contingent Objects

More influentially at the moment, there's debate among philosophers who accept quantified modal logic (and quantifying-in) about whether everything exists necessarily. In most (reasonably strong) quantified modal logics we can prove the following claim which seems to say that everything exists necessarily:[13]

$$(\forall x)\Box(\exists y)(y = x)$$

Hellman follows Kripke (Kripke 1963; Hellman 1994) in saying that familiar principles from sentential modal logic like the necessitation rule and K in S5[14] only apply to complete sentences in quantified modal logic. And perhaps this is intuitively motivated. We wouldn't want to say it's logically necessary or a tautology that $x = x$, because formulas with free variables aren't even sentences and thus lack truth-values.

However, this response is controversial. For example, an alternative approach would be to allow quantifying-in, but use free logic (Nolt 2018).[15] Also note that on

---

[13] In particular, if we take $\phi(x)$ to be $x = x \rightarrow (\exists y)(y = x)$ we easily see that $(\forall x)\phi(x)$ is logically true and thus infer $(\forall x)\Box\phi(x)$, i.e., $(\forall x)\Box[x = x \rightarrow (\exists y)(y = x)]$. We can thus infer the sentence above.

[14] These rules are, respectively: if $\vdash A$ then $\vdash \Box A)(\Box(A \rightarrow B) \rightarrow (\Box A \rightarrow \Box B)$.

[15] Switching to a free logic would let us block the above argument by blocking the initial proof that "$(\exists y)(y = x)$," rather than the application of necessitation to this formula in the last sentence, as free logics neither assume that all singular terms refer to members of the domain nor that the domain is non-empty.

I think this strategy is prima facie quite appealing, because it would allow us to capture the intuitive logical possibility of entirely empty domains. However, because as a matter of sociological fact, no free logic is currently widely accepted (and because avoiding quantifying-in makes the implications of adding any given axiom more obvious), I have preferred to sacrifice intuitions about empty domains

Kripke's approach (Kripke 1963), sentences like $(\exists x)\Diamond[\neg\text{Fox}(x) \wedge (\forall y)\text{Fox}(y)]$ are true if there are any contingent objects (a conclusion which can't be easily avoided[16]), a consequence which Williamson (2013) points out is fairly counterintuitive.[17]

3.3.1.3   Necessary Distinctness

Next, disagreement can arise about whether all pairs of things that are actually distinct are necessarily distinct. For example, Fine (2006) considers making this assumption and whether it can address Quinean worries about (logical) essences mentioned above:

There are, of course, familiar Quinean difficulties in making sense of first-order quantification into modal contexts when the modality is logical. Let me here just dogmatically assume that these difficulties may be overcome by allowing the logical modalities to "recognize" when two objects are or are not the same. Thus

$$\Box \forall x \Box (x = y \rightarrow \Box x = y)$$

and

$$\Box \forall x \Box (x \neq y \rightarrow \Box x \neq y)$$

will both be true though, given that the modalities are logical, it will be assumed that they are blind to any features of the objects besides their being the same or distinct.

But (to the extent that we have any grip on quantifying-in to logical possibility) this assumption is disputable. For example, some have argued that it's metaphysically (and hence presumably logically) possible for there to be two people who could have been one person. Suppose that two people are formed by a contingent event of a person splitting, e.g., a Star Trek transporter malfunction, or a brain getting split in half and each side regrowing. One might think these people are distinct but they could have been identical.[18]

3.3.1.4   Metaphysical Shyness?

Additionally, Linnebo (2018b) formulates a worry specific to Hellman, concerning the possibility of a kind of metaphysical or logical shyness. He writes, "Do we really know that there cannot be 'metaphysically shy' objects, which can live comfortably in universes of small infinite cardinalities, but which would rather go out of existence than to cohabit with a larger infinite number of objects?" This existence of such "shy" objects would pose a problem for Hellman, because it could block us from saying that

---

and use classical first-order logic rather than arguing for new views on both first-order logic and set theory in this book.

[16] Note that, when considering the truth-value of Fox(x) under an assignment of "x" to some contingent object o that doesn't exist at some possible world w, it seems we must say that x isn't in the extension of "Fox" at w, since it would be weird to insist that objects that don't exist at w were nonetheless foxes, and hence that $\neg Fox(x)$ should be true under this assignment.

[17] While this debate is commonly conducted in terms of metaphysical possibility, it naturally raises similar concerns for logical possibility.

[18] I take this point from Schwarz (2013).

every plurality of objects forming a hierarchy of a certain kind could be extended in a certain way.

Linnebo also notes that if Hellman's notion of logical possibility allows for an analog to metaphysically incompatible objects (e.g., two metaphysically possible knives formed by joining a single handle with different blades), this can make certain assumptions Hellman uses to justify the existence of Potentialist translations of ZFC come out false.

Paraphrasing sentences of set theory with modal sentences that quantify in to the ◊ of logical possibility forces us to consider when objects from one logically possible world are identical to, or counterparts of, objects in another. We are forced to ask whether, for some particular object, *that very object* could count as persisting in a world where the total universe has some cardinality, or some other possible object exists.

3.3.1.5   What to Do?

These controversies can raise doubts about whether our intuitions about quantifying-in are reliable[19] and whether we can choose axioms for modal logic which are both powerful enough to justify Potentialist formalizations of the ZFC axioms and clearly and (fairly) uncontroversially true, in the way we'd like foundational mathematical axioms to be.

One could debate about whether the disagreements above are best understood as a philosophical disagreement about a proposition (e.g., that everything exists necessarily) or as showing that we don't have a good grip on what quantifying-in means or that the formalism of quantified modal logic (that allows quantifying-in) means different things to different people. But, for my purposes, either option would be a sufficient reason to avoid formulating our foundational modal axioms (used to justify set theory) in terms of quantifying-in.

In general, one might try to solve this kind of problem by stipulating that sentences which quantify in to the ◊ should be understood as having whatever meaning is necessary to make certain axioms true. But note that, for the purpose of formalizing set theory (as evenly modestly truth-value realistically construed), this approach won't do. For insofar as we need there to be proof transcendent facts about set theory, we can't just say that any interpretation of our ◊ quantified modal statements that satisfies certain axioms (sufficient to justify Potentialist translations of set-theoretic claims) is equally intended. We have to try to latch on to an intuitively meaningful notion, about which truth can outrun proof.

Instead, I propose to solve the above problem in a different way: by eliminating quantification in to the ◊ of logical possibility, as we will see in Chapter 4. Beyond these philosophical motivations, in Chapter 5, I'll suggest this approach has certain practical advantages.

---

[19] My proposed account of set theory is compatible with taking Williamson to show that any modal notion *which allows quantifying-in* (such as metaphysical possibility) must have a fixed domain – provided one thinks it doesn't make sense to quantify in to logical possibility. Of course, it's not compatible with taking Williamson to show that every modal notion must have a fixed domain.

3.3.2   Justifying Replacement

Finally, issues about justifying Potentialism from an Actualist point of view remain. Merely adopting Hellman's Putnamian Potentialism doesn't suffice to secure our foundational aim of justifying set-theoretic theorems from obvious seeming assumptions.

Hellman does prove a version of the main theorem one needs to vindicate standard first-order reasoning about set theory (where $\phi^{\Diamond}$ denotes the potentialist translation of ϕ):

$$ZFC \vdash \phi \text{ then } \phi^{\Diamond}$$

However, the premises he uses in this proof don't (and aren't claimed to) seem clearly true. For instance, Hellman (1994) simply assumes the translation of Replacement into a Potentialist context as an axiom and explicitly flags that it is not intuitively obvious.

In later work, Hellman experiments with other justifications for Replacement. But it should be noted that in doing this his aim is only to motivate unifying principles by showing that key set-theoretic beliefs follow from a single natural hypothesis (as per Section 1.4), not significantly justify Replacement itself. So, the hypotheses from which Hellman derives Replacement don't seem any more clearly true than the Potentialist translation of Replacement, and often much less so. For example, Hellman (1994) considers a modal reflection principle, which would justify Potentialist Replacement but, just as in the Actualist case, seems no more obvious than Replacement itself,[20] and Roberts (2017) argues this principle is inconsistent with other axioms Hellman should plausibly endorse.

---

[20] Hellman motivates this principle by considering the following statement of a Potentialist Replacement principle: "The mathematical possibilities of ever larger structures are so vast as to be 'indescribable': whatever condition we attempt to lay down to characterize that vastness fails in the following sense: if indeed it is accurate regarding the possibilities of mathematical structures, it is also accurate regarding a mere segment of them, where such a segment can be taken as the domain of a single Structure." However, he notes this is inconsistent, and tries restricts its application to things consistent with ZFC_2. But this principle doesn't seem any more obvious than the reflection principles invoked by Actualists discussed in Section 2.4.

4   Overview of My Proposal

Let me now turn to my preferred form of Putnamian Potentialism, which will let us avoid the problem of controversies about quantified modal logic discussed in Chapter 3.

In this chapter, I'll clarify the notion of logical possibility in answer to the worries from Section 3.3.1 and explain how an independent line of reasoning from Boolos (1985) motivates accepting some such notion as a logical primitive. Then I'll introduce my key *conditional* logical possibility operator (which generalizes the logical possibility operator just mentioned) and discuss how using it to formulate Potentialist set theory is helpful.[1] However, I'll delay actually using this notion to paraphrase set-theoretic claims until Chapter 6.

4.1   What Is Logical Possibility?

4.1.1   On Logical Possibility

So, let me begin by clarifying and motivating the notion of logical possibility that I will appeal to (and one might think Hellman wants to invoke as well).

We seem to have an intuitive notion of logical possibility which applies to claims like $(\exists x)(\text{red}(x) \wedge \text{round}(x))$ and makes sentences like the following come out true:

- it is logically possible that $(\exists x)(\text{red}(x) \wedge \text{round}(x))$;
- it is not logically possible that $(\exists x)(\text{red}(x) \wedge \neg\text{red}(x))$;
- it is logically necessary that $(\forall x)(\text{red}(x)) \rightarrow \neg(\exists x)(\neg\text{red}(x))$.

This notion of logical possibility is interdefinable with validity. An argument is valid if and only if it's logically impossible for all its premises to be true and its conclusion to be false. And it is (roughly) what's analyzed by saying some theory has a set-theoretic model[2] (modulo concerns about size, as noted in Appendix F in the online appendix). It concerns whether some state of affairs is allowed by the most general "subject matter neutral" laws of how there can be some pattern of objects standing in relations of various arities (in something like Frege's sense of logical laws being subject matter neutral (Frege 1980)).

---

[1]   I first advocated doing this as a way to remove redundancies from Hellman's modal structuralism in Berry (2018a).

[2]   When considering non-first-order sentences we might specify that this model must treat all logical vocabulary standardly, so that, e.g., Henkin models of second-order quantification are not allowed.

Philosophers representing a range of different views of mathematics have made use of this notion and are comfortable applying it to non-first-order sentences.

To evaluate whether a claim ϕ is logically possible (in this sense), we hold fixed the operation of logical vocabulary (like $\exists$, $\wedge$, $\vee$, $\neg$) but abstract away from any further metaphysically necessary constraints on the application of particular relations. Thus, we consider all possible ways for relations to apply (including those ways that aren't definable). For example, it is logically possible that $(\exists x)(\text{raven}(x) \wedge \text{vegetable}(x))$, even if it would be metaphysically impossible for anything to be both a raven and a vegetable.

We also abstract away from constraints on the size of the universe, so that $\Diamond(\exists x)(\exists y)(\neg x = y)$ would be true even if the actual universe contained only a single object.

4.1.2 Contrast with Other Modal Notions

It may be useful to note how the above notion of logical possibility differs from three vaguely similar modal notions in the literature, namely Tarskian reinterpretability, metaphysical possibility and conceptual possibility.

The notion of logical possibility is (potentially) less demanding than the notion of truth under some Tarskian reinterpretation, for approximately the reason discussed above (and emphasized in Etchemendy (1990)). Certain scenarios might be genuinely logically possible but require the existence of more objects than actually exist, and hence not permit any Tarskian reinterpretation. For, Tarskian reinterpretations of a sentence must still take the sentence's quantifiers to range over some collection of objects in the actual world.

The notion of logical possibility is also prima facie less demanding than the notion of metaphysical possibility.[3] For, as Frege noted, the laws of logic hold at all possible worlds. Yet it would seem that statements like $(\exists x)(\text{raven}(x) \wedge \text{vegetable}(x))$ can require something which is logically possible but metaphysically impossible.

Finally, the notion of logical possibility is also less demanding than the notions of idealized conceivability and conceptual possibility at issue in debates over philosophical zombies and in Chalmers' *Constructing the World* (Chalmers 2012) (and are, inconveniently, sometimes also labeled logical possibility). For the notion of conceptual possibility reflects something like ideal a priori acceptability. So, when evaluating whether it is conceptually possible that ϕ we have to preserve all analytic truths associated with relations occurring in ϕ. In contrast (as I have noted above) logical possibility abstracts away from all such specific features of relations. Thus, for example, if we assume it is analytic that $(\forall x)(\text{bachelor}(x) \rightarrow \text{male}(x))$, then it will be logically possible but *not* conceptually possible that $(\exists x)(\text{bachelor}(x) \wedge \neg\text{male}(x))$.

---

[3] I want to leave it open the possibility that on some kind of ideal logical analysis, logical possibility turns out to be the same thing as metaphysical possibility. I'm just noting that we have a concept of logical possibility independent of this assumption, and that this suffices to give an attractive account of set theory.

4.1.3   Not Reducible to Set Theory

Because (as noted above) the notion of logical possibility is interdefinable with validity, I think nearly all my readers will accept that claims about logical possibility are meaningful.

However, at first glance, one might argue that claims about logical possibility are merely shorthand for claims about the existence of set-theoretic models. And if one identified logical possibility the notion of logical possibility with claims asserting the existence of set-theoretic models, then we'd have (at least) an uncomfortable regress, and one couldn't use the notion of logical possibility in formulating Potentialist set theory to solve the arbitrariness problem above.

Luckily, however, there are strong independent reasons for not doing this, pointed out in Boolos (1985), Gómez-Torrente (2000), Hanson (2006), Etchemendy (1990) and Field (2008). Many philosophers have argued, as follows, that we shouldn't identify claims about logical possibility with claims about set-theoretic models.

The claim that what's actual is logically possible is central to the notion of logical possibility (interdefinable with validity), if anything is. For an argument to be valid surely at least requires that it doesn't *actually* lead from truth to falsehood.

However, if we think about logical possibility in terms of set-theoretic models, then the actual world is strictly larger than the domain of any set-theoretic model (e.g., because it contains all the sets). So, it's not prima facie clear why we should assume that what can't be satisfied in any set-theoretic model isn't actually true. Thus, we seem to antecedently grip a notion of logical possibility (interdefinable with validity) on which it's an open question whether every logically possible state of affairs has a set-theoretic model.

Now it is *currently* possible for mathematicians talking about *first-order logical sentences* to replace talk of logical possibility with talk of set-theoretic models via the completeness theorem for first-order logic.[4] However, as Boolos puts it, "it is rather strange that appeal must apparently be made to one or another non-trivial result in order to establish what ought to be obvious: viz., that a sentence is true if it is valid" (Boolos 1985).

A further benefit of adopting a primitive logical possibility operator is that it lets us capture Boolos' intuition that there's something odd about identifying claims about logical possibility and validity with set-theoretic claims (claims about the existence of set-theoretic models). We can agree with Boolos that, "one really should not lose the sense that it is somewhat peculiar that if G is a logical truth, then the statement that G is a logical truth does not count as a logical truth, but only as a set-theoretical truth" and so reject cashing out claims about failures of logical truth/validity in terms of logically

---

[4] The completeness theorem shows that all syntactically consistent first-order theories have models. And the notion of logical possibility is intuitively "sandwiched between" syntactic consistency and having a model (anything that has a model must be logically possible, and anything that's logically possible must be syntactically consistent), so this shows that all three notions apply to exactly the same first-order logical sentences (Field 2008).

contingent claims about the existence of certain objects (even mathematical objects). To foreshadow slightly, following Boolos' suggestion, I will treat the ◊ of logical possibility a primitive modal operator, and furthermore *logical* operator whose meaning must be held fixed when we're evaluating claims about logical possibility and entailment. Thus, we can affirm that facts about logical possibility are themselves logically necessary truths.

In view of all the points above, I take it that there's no problem in (and indeed significant independent motivation for) accepting that we have a primitive modal notion of logical possibility. Talk of arguments' validity (in some sense) seems to be widely understood and useful. And cashing validity claims out by appeal to a primitive modal notion of logical possibility (rather than attempting to reduce it to a notion of having a set-theoretic model or truth under some Tarskian reinterpretation) seems like the wisest course.

4.2   Conditional Logical Possibility

So much for clarifying and defending appeal to the logical possibility operator. Now let's turn to the philosophical controversies about quantified modal logic (and the practical problems of making axioms for it surveyable) that threatened to block our foundational ambitions in Chapter 3. I propose that we can solve these problems by thinking about Potentialist extensibility claims as concerning what's allowed by a given structure, rather than what's possible for given objects. And I'll suggest that a certain natural generalization of the logical possibility operator will help us do this.

4.2.1   Motivation

As modal Structuralists like Hellman have observed, mathematicians are unconcerned with questions about the nature and essence of particular objects. They don't care whether the number "1" refers to the set or the set, or Julius Caesar, only that whatever objects the predicate "natural number" applies to have a certain structure (under whatever relations are expressed by the terms "successor," " +," "." etc.). And any copy of this structure (whether formed of sets or emperors) is, in some sense, equally relevant to number theory.[5] Considering any objects under any relations of the right arity will do, provided the right pattern in how these relations apply is instantiated. Neither the particular relation playing the role of "successor" nor the particular objects playing the role of "numbers" matter, from a pure mathematical point of view.

---

[5] So, for example, suppose you take some strokes that form an instance of the natural number structure under "to the right of" and erase one stroke and then rewrite a new stroke in the same place (so that the patterns of how the relations "stroke" and "to the right of" apply is preserved but the objects are different). Then you have another copy of the natural number structure and (in a way) nothing that matters has changed. Similarly the particular relations under which objects form a copy of the natural number structure don't matter (turning your stroke sequence on its side and changing each stroke to an exclamation mark so that you now have an ω sequence of exclamation marks under "below" produces something equally relevant). Any relations of the right arity will do.

Developing Potentialist set theory requires us to compare possible structures, to give some precise meaning to claims about how it would be possible for one initial segment of a cumulative hierarchy of sets to extend another. And Putnam and Hellman do this by quantifying-in. But perhaps the point that mathematics is fundamentally concerned with structure alone rather than objects suggests a different way to achieve the same goal.

I'll suggest it suffices to reconstruct Potentialist set theory to consider what's possible given the *pattern* of how some relations (instantiating some mathematically relevant structure) apply, rather than asking what's possible for the particular objects or plurality of objects which these relations happen to apply to. As the discussion of metaphysically shy objects in Section 3.3.1.4 suggests, it doesn't intuitively matter to the truth-value of a Potentialist set-theoretic claim whether some particular objects forming an iterative hierarchy structure (under some relation like "there is an arrow pointing from ... to ...") could *continue to exist* while this structure is supplemented by additional objects so as to form an extending iterative hierarchy. All that matters is whether the *structure* of how the relation "there is an arrow pointing from ... to ..." applies to these objects could be preserved, while objects forming a suitable extended hierarchy (under some other relation) are added.

4.2.2   Introducing Conditional Logical Possibility

To informally introduce the notion of conditional logical possibility (aka logical possibility given structural facts about how some relations apply) consider claims that some map isn't three colorable.

When you say a map isn't three colorable, you don't just mean that it would be physically or metaphysically impossible for the map to be three colored (without some change in the extensions of "country on the map" and "adjacent to"). Rather, you are saying something stronger, which we might make explicit by saying that it's *logically impossible* given the *structural facts* about (aka pattern of) how the relations "country" and "adjacent to" apply for the map to be three colored. This means two things.

First, the *mere pattern* of how the relations "country" and "adjacent" apply (rather than any special features of the objects these relations apply to) prevent the map from being three colored. So, for instance, if wars and revolutions change replace one country with another and shift national boundaries but *don't* change the pattern of how countries are related by adjacency then this new map is three colorable only if the old one was, as it's the structure of how the relations "adjacent" and "country" apply which determines if the map is three-colorable.

Second, this pattern of how the relations "country" and "adjacent" apply makes three coloring *logically* (as opposed to merely physically or metaphysically) impossible, i.e., it blocks three coloring in virtue of completely general, subject matter neutral, laws that treat all relations of the same arity alike. Thus, it's equally impossible for the map to be three scented or three textured. And if any other relations (e.g., "city" and "has a direct flight to") instantiated the same pattern, then they wouldn't/couldn't be three colored/textured etc. either.

The notion of conditional possibility ($\lozenge_{\ldots}$) generalizes the notion of logical possibility ($\lozenge$) in a way that lets us naturally express claims like the three colorability statement above. The subscript will specify certain relations – in this case "is a country" and "is adjacent to" – whose pattern of application we want to hold fixed. And, as will become clear in a moment, we can write the non-three-colorability claim above as follows:

Non-Three-Colorability: $\neg \lozenge_{\text{adjacent,country}}$ Each country is either yellow, green or blue and no two adjacent countries are both yellow, both blue or both green.

I will read this as meaning, "It's not logically possible, given the structural facts about how 'adjacent' and 'country' apply, that: each country is either yellow, green or blue and no two adjacent countries are the same color."

If you accept a primitive modal notion of logical possibility (interdefinable with validity) advocated in Section 3.1, it seems only natural to allow restriction of that notion to the scenarios which preserve the structure of how some relations apply. To further precisify what I mean, consider the following, even simpler, example:

Crowded Cats: Given what cats and baskets there are, it is logically impossible that each cat is sleeping in a different basket.

If we take logical possibility to mean logical possibility simpliciter, this sentence must be false. However, it also has an intuitive reading on which it could be true. One might express the latter by saying "Cathood and baskethood apply in a way that ensures that (as a matter of mere logic and combinatorics) it can't be that each cat is sleeping in a different basket." A moment's thought will reveal that (on this reading) the above sentence is true if and only if there are more cats than baskets.

As we saw above, I will express such claims about conditional logical possibility using an operator $\lozenge_{(\ldots)}(\ldots)$. This conditional logical possibility operator takes a sentence ϕ and a finite (potentially empty) list of relation symbols $R_1, \ldots, R_n$ and produces a sentence $\lozenge_{R_1,\ldots,R_n}$ which says that it is logically possible for ϕ to be true, without any change to (structural facts about) how the relations $R_1, \ldots, R_n$ apply. But for ease of reading, I will sink the specification of relevant relations into the subscript as follows: $\lozenge_{R_1,\ldots,R_n}$. So, I'll write the claim about cats and baskets above as follows:

Crowded Cats: $\neg \lozenge_{cat,basket}$ [Each cat slept in a different basket.]

Now let me specify three things about how this notion of conditional logical possibility is to be understood. The first concerns how conditional logical possibility relates to logical possibility simpliciter. We saw that claims about logical possibility simpliciter ($\lozenge$) concern what's possible if we let both the size of the domain of discourse and the application of relations to that domain vary with complete freedom. In contrast, claims about conditional logical possibility ($\lozenge_{R_1,\ldots,R_n}$) concern what's logical possible if we hold fixed the structural facts about how some relations $R_1, \ldots, R_n$ apply (while still letting the size of the domain extending this structure and the application of other relations vary freely).

Second, what does it mean to "hold the (structural facts) about how some relations apply fixed?" In line with the motivating case above, keeping the structural facts about how some relations apply fixed doesn't mean preserving these relations' extensions (the particular objects they apply to/relate). Rather it means preserving the pattern of how all these relations apply. So, for example, metaphysically possible scenarios where one cat dies early and one kitten is born early will count as preserving the *structural facts* about what cats and baskets there are (i.e., the pattern formed by how cathood and baskethood apply). And preserving the structural facts about how cat($\cdot$) and basket($\cdot$) will require preserving: the number of cats, the number of baskets and the number of things (0) that are both cats and baskets. In more familiar Platonist language, we might say it means holding the extensions (where these are n-tuples for n-ary relations) of these relations fixed *up to isomorphism*.

To bring out the difference between preserving structure and preserving objects at issue, and explain why claims formulated in terms of structure preserving logical possibility should be vastly less controversial than claims about *de re* possibility note that I can suspend judgement (or deny that there's a legitimate question) about which properties Nixon had essentially (politician, human, liar, man), while accepting and evaluating claims about what's metaphysically or logically possible given the structure of how relations like "reports to" and "is a politician" apply.

Third, note that I don't take structure preservation to require holding fixed the whole size of the universe. The structure which $\Diamond_{R_1,\dots,R_n}$ claims holds fixed is the structure formed by the objects which at least one of the relations $R_1,\dots,R_n$ apply to, considered under the relations $R_1,\dots,R_n$. In this case that means considering the structure of the cats and baskets under the relations cat($\cdot$) or basket($\cdot$).[6]

To motivate this way of thinking about what it takes to preserve/agree on structural facts about some list of relations, consider when we'd say two different interpretations of some person's language agree on the *structure* of the natural numbers (under successor). Two interpretations will agree on the structure of the natural numbers if they both take "number" and "successor" to apply to some ω sequence – even if they disagree about the total size of the universe or whether Julius Caesar or the empty set are identical to any numbers etc.[7] My understanding of what it takes to keep structural facts fixed generalizes this way of thinking about what's required to preserve the natural number structure (the structure of objects under the relations "natural number" and "successor").

---

[6] Speaking metaphorically, we want to consider logically possible scenarios where there's a bijection f between the set of objects which at least one of the relations $R_1,\dots,R_n$ apply to in the actual world, and the set of objects these relations apply to in that logically possible situation, such that $R(x,y)$ iff $R(f(x),f(y))$.

[7] Or consider the way that a Platonist would say the structure of the natural numbers is fixed necessarily and will always remain the same, even if the total size of the universe can be changed by the creation or destruction of physical objects or changes to the structure of space etc.

4.2.3   Isomorphisms and Kripke Models

Given readers' presumed prior familiarity with set theory and metaphysics, it may help indicate the modal notion I have in mind to relate conditional logical possibility facts to common ideas about set theory and possible worlds. However, it should be noted that this comparison is made purely for expository efficiency. I'm putting conditional logical possibility forward as a conceptual and metaphysical primitive which we *could* learn by immersion, in the same way we learn "set" and " $\in$ " and all I aim to do here is to evoke the idea using more familiar notions.

   If we could talk about functions between (the objects in) different logically possible worlds, then we could specify what it takes to hold the structural facts about how some relation (say, "admires()") applies fixed, in terms of isomorphisms as follows.

   A world w_2 counts as holding fixed the structural facts about how "admires" applies in w_1 iff the objects related by admiration w_1 are isomorphic to those related by admiration in w_2 (you can map one collection of objects to the other in a way that's 1-1 and respects admiration):

A logically possible world w_2 counts as holding fixed the structural facts about how admires() applies in w_1 iff some function f bijectively maps the objects which either admire or are admired in w_1 to the objects which either admire or are admired in w_2, so that for all objects x and y in w_1 which either admire or are admired in w_1, we have x admires y iff $f(x)$ admires $f(y)$.[8]

We will also see that facts about Potentialist set theory can be mimicked (modulo limitations of size) by talk about models in set theory with ur-elements, in Section 4.2.6 below.

   In terms of Kripke models logical possibility simpliciter demands that every world be accessible to every other world. Indeed, logical possibility simpliciter satisfies the modal axiom system S5. One can think of conditional logical possibility as modifying this picture by allowing the selection of an accessibility relation based on which relations are subscripted. Thus, the $\Diamond_{R_1,\ldots,R_n}$ operator considers the same set of possible worlds but makes use of an accessibility relation in which w_2 is accessible to w_1 iff there's a bijection between the objects at w_1 and w_2 which fall under some R_i that respects the relations $R_1, \ldots, R_n$. As the accessibility relation for $\Diamond_{R_1,\ldots,R_n}$ is an equivalence relation, $\Diamond_{R_1,\ldots,R_n}$ will also satisfy S5.

4.2.4   Comparison with Shapiro

One can further explain and motivate the notion of conditional (i.e., structure preserving) logical possibility by relating it to Stewart Shapiro's notion of structures qua abstract objects, instantiated by various physical (and otherwise) systems in Shapiro (1997).

---

[8] More generally, a logically possible world w_2 preserves the structural facts about how relations $R_1, \ldots, R_n$ (say admires() and cat()) apply iff some function f bijectively maps the objects which $R_1, \ldots, R_n$ apply to in w_1 (i.e., those things which are either cats or admire something or are admired by something) to the objects which $R_1, \ldots, R_n$ apply to in w_2 in a way that respects all these relations.

There he says that a structure is "the abstract form" of a system of objects, which we get by "highlighting the interrelationships among the objects and ignoring any features of them that do not affect how they relate to other objects in the system." Thus, for example, "The natural-number structure is exemplified by the strings on a finite alphabet in lexical order, an infinite sequence of strokes, an infinite sequence of distinct moments of time, and so on." Adding or subtracting objects to the world outside of a given system, will make no difference to which structure that system instantiates.

Again, I mean to propose the conditional logical possibility operator as a conceptual primitive, and I don't endorse Shapiro's structures qua abstract objects.[9] However, one can (roughly) explain my notion of conditional logical possibility (aka structure preserving logical possibility) in terms of Shapiro's notions as follows.

It is logically possible, given the $R_1, \ldots, R_n$ facts, that ϕ, i.e., $\Diamond_{R_1, \ldots, R_n}$ iff some logically possible scenario makes ϕ true while holding fixed what structure (in Shapiro's sense) the system formed by the objects related by $R_1, \ldots, R_n$ (considered under the relations $R_1, \ldots, R_n$) instantiates.

4.2.5   Nested Logical Possibility Claims

If we accept the notion of conditional logical possibility, we can also make *nested* logical possibility claims. That is, we can make claims about the logical possibility of scenarios which are themselves described in terms of logical possibility. So, for example, I could say that it would be logically possible for the Crowded Cats claim above to be true.

Possibly Crowded Cats: $\Diamond (\neg \Diamond_{\text{cat,basket}}[\text{Each cat slept in a different basket.}])$

When evaluating such sentences, I want to hold fixed the meaning of the conditional logical possibility operator (i.e., treat it as a piece of logical vocabulary). And I take the above sentence, Possibly Crowded Cats, to express a truth because (reading from the outside in):

- it is logically possible (holding fixed nothing) that there are four cats and three baskets;
- relative to the logically possible scenario where there are four cats and three baskets, it is not logically possible (given what cats and baskets there are), that each cat slept in a basket and no two cats slept in the same basket.

Note that when evaluating a nested logical possibility claims with the form $\Diamond (\neg \Diamond_R \psi)$, I will take the subscripted $\Diamond_R$ to freeze the facts about how the relation R applies in the state of affairs under consideration, which may *not* be the state of affairs in the actual world.[10]

---

[9] See Section 17.2 for some reasons why.

[10] So, for example, $\Diamond$CATS expresses a metaphysically necessary truth. For, whatever the actual world is like, it will always be logically possible for there to be, say, three cats and two baskets. And any such scenario is one in which it is logically necessary (holding fixed the structural facts about what cats and baskets there are) that: if each cat slept in a basket, then multiple cats slept in the same basket. So, it is metaphysically necessary that $\Diamond$CATS even if the actual world contains more baskets than cats.

In this way I take logical possibility sentences of the form $\Diamond_{R_1,\ldots,R_n}\phi$ to be meaningful, even in cases where ϕ itself makes claims about conditional logical possibility. I will work in a formal language $\mathcal{L}_\Diamond$, which I will call the language of logical possibility, that allows such claims.[11] However, as foreshadowed above, the language of logical possibility will not include sentences which quantify in to the $\Diamond$ of logical possibility, e.g., sentences of the form $(\exists x)\Diamond\phi(x)$.

4.2.6   Comparison to Set Theory with Ur-Elements

We can use the familiar background of Actualist set theory to *mimic* the intended truth conditions for statements in a language containing the logical possibility operator $\Diamond$ alongside usual first-order logical vocabulary (where distinct relation symbols R_1 and R_2 always express distinct relations) as follows:

A formula ψ is true relative to a model $\mathcal{M}$ ($\mathcal{M} \models \psi$) and an assignment ρ which takes the free variables in ψ to elements in the domain of $\mathcal{M}$[12] just if:

- $\psi = R_n^k(x_1,\ldots,x_k)$ and $\mathcal{M} \models R_n^k(\rho(x_1),\ldots,\rho(x_k))$;
- $\psi = x = y$ and $\rho(x) = \rho(y)$;
- $\psi = \neg\phi$ and ϕ is not true relative to $\mathcal{M}, \rho$;
- $\psi = \phi \wedge \psi$ and both ϕ and ψ are true relative to $\mathcal{M}, \rho$;
- $\psi = \phi \vee \psi$ and either ϕ or ψ are true relative to $\mathcal{M}, \rho$;
- $\psi = \exists x \, \phi\,(x)$ and there is an assignment ρ' which extends ρ by assigning a value to an additional variable v not in ϕ and $\phi[x/v]$ is true relative to $\mathcal{M}, \rho'$;[13]
- $\psi = \Diamond_{R_1,\ldots,R_n}\phi$ and there is another model $\mathcal{M}'$ which assigns the same tuples to the extensions of $R_1,\ldots,R_n$ as $\mathcal{M}$ and $\mathcal{M}' \models \phi$.[14]

Note that this means that $\bot$ is not true relative to any model $\mathcal{M}$ and assignment ρ.

If we ignore the possibility of sentences which demand something coherent but fail to have set models because their truth would require the existence of too many objects, we could then characterize the true sentences in the language of logical possibility as follows:

Set-theoretic Approximation: A sentence in the language of logical possibility is true (on some interpretation of the quantifier and atomic relation symbols of the language of logical possibility) iff it is true relative to a set-theoretic model whose domain and extensions for atomic relations captures what objects there are and how these atomic relations actually apply (according to this interpretation) and the empty assignment function ρ.

---

[11] To describe this language more explicitly, fix some infinite collection of variables and relation symbols of every arity together with $\bot$ and define the language of logical possibility to be the smallest language built from these variables using these relation symbols and equality closed under applications of the normal first-order connectives and quantifiers and $\Diamond_{\ldots}$ (where $\Diamond_{\ldots}$ expressions can only be applied to sentences (so there is no quantifying-in). We will also use $\Box_{\ldots}$ in our sentences but regard it as an abbreviation for $\neg\Diamond_{\ldots}\neg$.

[12] Specifically: a partial function ρ from the collection of variables in the language of logical possibility to objects in $\mathcal{M}$, such that the domain of ρ is finite and includes (at least) all free variables in ψ.

[13] As usual, $\phi[x/v]$ substitutes v for x everywhere where x occurs free in ϕ.

[14] As usual, I am taking $\Box$ to abbreviate $\neg\Diamond\neg$.

4.3   Advantages

4.3.1   Some Advantages

Appealing to the notion of conditional logical possibility when formulating Potentialist set theory provides three important benefits.

First, of course, it lets you do Potentialist set theory without incurring controversial commitments to object essences and cross-world identity facts as discussed in Section 3.3.1. So, it lets you assert axioms which can be accepted by those who share Quine's doubts about whether quantifying-in is meaningful and avoid Linnebo's shyness worry.[15]

Second, even if you don't object to the metaphysical primitives required to make sense of quantifying-in (and feel that most relevant disagreement could be revealed to be mere verbal disputes by more carefully evoking Hellman's intended reading of quantifying-in) stating axioms in terms of conditional logical possibility helps us articulate principles that can be widely and easily recognized as true. Of course, this is not to say that people can't *philosophically* disagree about the meaning of the conditional logical possibility operator,[16] or that disagreement over mathematical axioms is completely impossible. However, there aren't multiple widely held views about the nature of logical possibility which would assign different truth-values to commonly used sentences.

Third, there's a practical benefit to using the conditional logical possibility operator rather than quantifying-in because it (in effect) cleaves good reasoning about logical possibility into two parts:

- in one part we use standard first-order logic to reason about a given logically possible scenario;
- in another part, we use special modal-structural principles to establish which scenarios are logically possible, and "transfer" facts about one scenario to another.

This helps us avoid the potentially confusing and hard to survey interactions between modal principles and free variables that we see in examples like the proof of the

---

[15]   Note that one might well accept that there are definite facts about metaphysical or logical possibility such as could be "coded" by set-theoretic models specifying the size of the domain and extensions for properties, while not thinking there are meaningful (or non-context relative) facts about essences. I can specify the facts about what metaphysically possible worlds there are in terms of how many objects exist in each and how all properties apply (hence pining down facts about what's conditionally logically possible with respect to each possible world w and list of relations $R_1, \ldots, R_n$), without telling you anything about essences or counterpart hood relations which would let you determine facts about what's *de re* metaphysically possible for a given individual. Thus, one might well think it's meaningful to ask whether it's structure-preservingly possible that ϕ without asking whether certain particular objects could exist in a world where ϕ.

[16]   There are plenty of ways of disagreeing about how the intuitive notion of conditional logical possibility should be cashed out, e.g., disagreements about whether logical possibility simpliciter should be understood in terms of possible interpretations for words transfer to this case.

converse Barcan-Marcus formula. This is an especially important property for foundational axioms to have, as their truth should be evident.

4.3.2 Simplification

A final advantage of reformulating Potentialist set theory using conditional logical possibility is that it lets us eliminate appeals to second-order or plural quantification from our Potentialist paraphrases. Hellman uses these notions to give categorical descriptions of the natural numbers and (categorical up to height) of the iterative hierarchy of sets. However, we can use the conditional logical possibility operator to do this work (Berry 2018a), thereby simplifying of our basic ideology.

4.3.2.1 Natural Numbers

First, consider the natural number structure. We can use conditional logical possibility to state axioms that uniquely pin down the intended structure of the natural numbers (a job which philosophers usually use second-order quantification to do) as follows. Note that the second-order induction axiom below, when combined with first-order axioms, suffices to pin down a unique model of the natural numbers (up to isomorphism):

$$(\forall X)\Big[\Big(X(0) \land (\forall n)\big(X(n) \rightarrow X(n+1)\big)\Big) \rightarrow (\forall n)\big(X(n)\big)\Big]$$

We can reformulate this claim using conditional logical possibility as follows:[17]

- **Induct:** $\Box_{\mathbb{N},S}$ If 0 is happy and the successor of every happy number is happy then every number is happy.

In other words: it is logically necessary, given how $\mathbb{N}$ and S apply, that if 0 is happy and the successor of every happy number is happy then every number is happy. Note that this has the same force as the second-order quantifier, which can pick out a counterinductive (i.e., a successor closed collection of natural numbers which doesn't contain every natural number) collection of natural numbers if and only if it would be logically possible for happy to apply to exactly those numbers, in violation of the above principle.

Thus, we can give a categorical description, $PA_\Diamond$, of the natural numbers in terms of conditional logical possibility by conjoining the above induction axiom with the familiar first-order axioms of PA^- (i.e., all the axioms of Peano Arithmetic except for the induction axioms). Recall that the second-order Peano Axioms are just the axioms of PA^- plus the second-order induction axiom above.[18]

---

[17] I write 0 below for readability but recall that one can contextually define away all uses of 0 by instead using its unique characterization as the unique element which isn't a successor.

[18] See Section J.3 of the online appendix for a more formal presentation of this point.

4.3.2.2   IHW

We can use essentially the same trick to eliminate second-order quantification to characterize the iterative hierarchy conception of the sets as spelled out in Definition 2.1.

There are two difficulties we face in spelling out this conception: capturing the notion of a well-ordering and ensuring the full-width requirement (i.e., that at any level l there are sets corresponding to all ways of choosing a collection of sets available at levels below l). I will only discuss the second concern here, as capturing the notion of a well-ordering is a simple modification of the mechanism we used to construct $PA_{\Diamond}$, and those interested in the details can consult Definition E.2 in Section E of the online appendix.

When it comes to the full-width requirement, we can again use conditional logical possibility to substitute for second-order logic. One expresses this idea in second-order logic by saying that, for every way a second-order object X can apply to the sets available below l, there's a set available at level l which contains exactly the objects that X *applies to*. But we can express the same idea by saying it's necessary, holding fixed the structure of the sets, that any way the one-place predicate happy applies to the sets available below l corresponds to the membership of some set available at level l (see Definition A.2 in Appendix A).

4.3.2.3   Avoiding Duplication of Primitives

I would further suggest that eliminating second-order and plural quantification from our formulations of Potentialist set theory in this way allows us to avoid a kind of unappealing conceptual redundancy.

Specifically, I think there's a kind of undesirable conceptual duplication in employing both the $\Diamond$ of logical possibility and a notion of second-order or plural quantification treated as an unrelated primitive (as Hellman does). For, intuitively, there's something in common between the way we consider "all possibilities" for how some first-order relations could apply when evaluating logical possibility and the way we consider "all possibilities" for choosing some first-order objects from a given collection when considering what second-order objects exist or when employing plural quantification.

Perhaps Actualists about set theory can straightforwardly explain this similarity. For they can define both notions in terms of what sets exist.[19] In particular, they can appeal to the same notion of "all subsets over a given first-order domain" when defining logical possibility in terms of the existence of a set-theoretic model and when cashing out second-order (and perhaps plural) quantification in terms of set existence.

But Putnamian Potentialists cannot do the same. For, we Putnamian Potentialists understand set existence in terms of possibility, rather than the other way around. So, we can't account for the sense of conceptual overlap between the notions of logical possibility and second-order quantification by cashing out both notions in terms of set theory. Thus, we lose the above benefit and, e.g., Hellman winds up treating logical

---

[19] Or at least, they can do this if we bracket Field's objection to identifying claims about logical possibility with claims about set theory discussed in Section 4.1.

possibility and second-order logic/plural quantification as separate conceptual primitives.

Happily, however, we can solve this problem if we embrace the notion of conditional logical possibility, for this single notion can be used to articulate and analyze both claims about logical possibility and second-order quantification.[20]

4.3.3   Ontology and Conceptual Primitives

In this chapter, I have proposed that we should make one choice of primitive (at least for a certain foundational project), while another has been historically familiar. Some readers may fear that employing a conditional logical possibility operator is, or enables, cheating at the project of ontology.

However, it should be noted that my aim is not to defend any kind of materialism or Nominalism (I'll ultimately argue for the existence of some pure mathematical objects). Nor do I mean to argue that facts about set theory are somehow metaphysically or epistemically trivial (or in any other sense a "free lunch").[21] For example, I take the concept of logical possibility to be a significant part of fundamental ideology, something that should certainly be counted when evaluating the metaphysical parsimony of any theory that employs it.

I'll suggest that philosophy of set theory will go better in certain ways (e.g., we can do a better job avoiding intuitive paradoxes and explaining why apparently good mathematical arguments are justified) if set theory is formulated potentialistically, using the conditional logical possibility operator indicated above. I ask that readers evaluate this choice of primitives on the basis of philosophical fruitfulness, problems raised and solved, and avoidance of redundancy, rather than by familiarity bias. If taking conditional logical possibility as a primitive is favored on the former grounds, then (I take it) using it as a primitive when formalizing set theory is (in some sense) appropriate and acceptable.

4.4   Looking Forward

In Chapter 6 I will provide details about how to use (nested) conditional logical possibility claims to state Putnamian Potentialist paraphrases of set theory which avoid quantifying-in (and plural and second-order quantification).

Following Hellman, my paraphrases will specifically invoke a notion of logical possibility (so metaphysical limits on the cardinality of objects are irrelevant), and I will develop Potentialism without claiming that some Actualist perspective on set theory is equally good. But, following Putnam, I will employ non-mathematical

---

[20] We have seen how to do this in for the purposes of second-order claims needed to formulate Potentialist set theory. See Berry (2018a) for an argument that we can reformulate second-order claims more generally

[21] I will consider what a Nominalist who thinks logical possibility facts are ontologically innocent can say about indispensability worries in Part II.

relations of suitable arities (e.g., "pencil point" and "an arrow from ... to ..." or "angel" and "admires") to talk about the possible existence of iterative hierarchy structures, rather than second-order class and function objects (or pluralities simulating such objects). I will also differ from Hellman in considering iterative hierarchies satisfying a version of IHW rather than hierarchies satisfying ZFC_2.

5 Parsonian Potentialism

5.1 Introduction

I will conclude Part I of this book by contrasting the Putnamian school of Potentialism to be developed in the rest of this book with a different, Parsonian, approach developed by Parsons (1977, 2005, 2007), Linnebo (2010, 2013b, 2018a), Roberts (2017, 2018) and Studd (2019). In a nutshell, the difference between the Parsonian and Putnamian schools is that Parsonians interpret set theory as talking about what *sets* (as objects with a special kind of essence) could be formed, while Putnamians understand set theory as making claims about how structures satisfying an explicit axiomatization for an initial segment of the set-theoretic hierarchy (e.g., ZFC_2 or IHW_2) could be extended.

At first glance, the choice between Parsonian and Putnamian approaches to set theory makes little difference to our foundational project. Advocates of both approaches have proved that their favored Potentialist translations of all theorems of ZFC are provable from certain modal principles, but don't claim prima facie obviousness for (all) these modal principles. Indeed, both sides tend to, in effect, take their Potentialist translation of Replacement as an axiom (sometimes noting that similar assumptions have been made elsewhere (Linnebo 2013)). However, we'll see that it turns out to be more convenient to adopt a Putnamian framework (at least temporarily) for this justificatory project.

5.2 The Parsonian Approach

Linnebo (2018b) explains the contrast between his preferred Parsonian approach to Potentialist set theory and the Putnamian Potentialism discussed in Chapter 3:

[On a Parsonian approach to set theory] the idea is not to "trade in" one's mathematical objects in favor of modal claims about possible realizations of structures but rather to locate some modally characterized features in the mathematical objects themselves. The mathematical universe is not "flat." Rather, some of its objects stand in relations of ontological dependence, and the existence of some of its objects is merely potential relative to that of others.

A multiplicity of objects that exist together can constitute a set, but it is not necessary that they do. Given the elements of a set, it is not necessary that the set exists together with them. ... However, the converse does hold and is expressed by the principle that the existence of a set implies that of all its elements. (Parsons, 1977: 293–4)

As Parsons emphasizes, this approach can also be used to explicate the influential iterative conception of sets, which tends to be explained by suggestive but loose talk about a "process" of "set formation." It would be better, Parsons claims, to replace this talk of time and construction with "the more bloodless language of potentiality and actuality."

So, the Parsonian takes the term "set" to have pre-existing meaning (and facts about the essential nature of sets to do critical work in their theory), while (as we have seen) the term "set" is completely eliminable from the Putnamian's theory. And Parsonian Potentialists take facts about what pure sets exist to be (in some sense of the word) contingent, with the existence of a set requiring the existence of that set's elements, but the overall height of the hierarchy of sets being contingent. Accordingly, Parsonian paraphrases of set-theoretic sentences have a similar large-scale structure to Putnamian paraphrases, replacing $\exists$ claims with $\Diamond$ claims and $\forall$ claims with $\Box$ claims. However, they take the relevant notion of possibility to concern what sets could (in some relevant sense) be formed. Also, Parsonians don't write any description of the iterative hierarchy structure into their Potentialist paraphrases. Instead, they take the fact that whatever sets exist must form (part of) an iterative hierarchy to fall out of – and be explained by – facts about the essences of sets and dependency relations between them.

For example, we saw that a Putnamian like Hellman might paraphrase "$(\forall x)(\exists y)(x \in y)$," as follows:[1]

$$(\forall V_1)(\forall x)[x \in V_1 \rightarrow \Diamond(\exists V_2)(\exists y)(y \in V_2 \wedge V_2 \geq V_1, \wedge x \in y)]$$

If we were to fully expand out the notation above, the resulting sentence would only use modal and logical primitives (not including either set or $\in$).

In contrast, Parsonians would translate "$(\forall x)(\exists y)(x \in y)$" more simply along the following lines:

$$(\forall x)[\text{set}(x) \rightarrow \Diamond(\exists y)(\text{set}(y) \wedge x \in y)]$$

They'd then appeal to substantive assumptions about set essences and what they entail about the possibility of set formation. For example, Linnebo and Studd take the fact that whatever sets have been formed always fit into an iterative hierarchy to be explained by facts about sets and plurals like the following:

- There are pluralities xx corresponding to (so to speak) all possible ways of choosing some objects that already exist (e.g., some sets that have already been formed)
- Whenever there's a plurality xx of sets, a corresponding set (i.e., a set whose elements are exactly the members of the plurality) could be formed.
- Sets and pluralities have their elements necessarily (so a set can't be formed before its elements have been formed), and sets are extensional (i.e., two sets are identical iff they have the same elements).

---

[1] Recall that here we are using quantification over all V_i as shorthand for quantification over all second-order objects X, f (or pluralities simulating them) satisfying some axioms like ZFC$_2$ (in the sense that $C_2[\text{set}/X, \in/f]$).

Thus, we could imagine a Parsonian hierarchy of sets growing as follows (if we knew what forming a set involved). The empty plurality always exists. So, an empty set could be formed. Form it. Now there's a plurality xx whose sole member is the empty set, so a set | could be formed. Form that. Now that both these sets exist, there are four pluralities xx of sets, and two of them correspond to sets we don't already have. So, we could form $\{\{\{\}\}\}$ and $\{\{\{\}\}, \{\}\}$, etc.

Remember, however, there are two readings of set-theoretic talk. In especially literal philosophical contexts, like the paragraph above, we can quantify over the sets that literally exist. However, in mathematical contexts, talk which appears to say that certain sets exist is always shorthand for corresponding claims about what sets could be formed.

5.2.1 Which Modal Notion?

One obvious question (and potential source of problems for the Parsonian) is this: how shall we understand the Parsonian's modal notion ◊? In what sense *could* there have been a different number of (pure) sets? And how many sets are there really? For example, if we understand talk of possible set formation as making a claim about how one could reconceptualize the world to think in terms of more sets (as we will see Linnebo does in (Linnebo 2018a)), how many sets are mathematicians *currently* thinking and talking in terms of? It's prima facie unclear how the Parsonians can answer this question in a principled fashion (especially if mathematical practice is always better understood by interpreting mathematicians as thinking potentialistically).

Parsons (1977) argues that we can't understand the possibility invoked in Parsonian paraphrases as meaning physical, metaphysical, mathematical or logical possibility as follows. One can't appeal to physical or metaphysical possibility, because the existence of sets isn't physically or metaphysically contingent. Similarly, Parsons understands mathematical possibility to mean possibility dropping "all constraints of a metaphysical nature" and considering only what is "compatible with the laws of mathematics" (where the latter include facts about what sets exist). Thus, he also holds that it wouldn't be mathematically possible for there to be a larger/smaller set-theoretic universe.

What about logical possibility? Linnebo (2018b) notes that appeal to "'logical modality in the strict sense' . . . is fairly quickly set aside by Parsons, who finds it to be 'either . . . an awkward notion generally or not in the end [different] from mathematical modality'." Now I take the arguments of Section 4.1 to show that there is a very natural and appealing notion of logical possibility (interdefinable with validity). However, we cannot interpret the ◇ occurring in Parsonian formalization of set theory to mean logical possibility in this sense. For key claims that the Parsonian wants to say are necessary (e.g., the fact that the sets are extensional) aren't logically necessary.[2]

So, what modal notion should the Parsonian invoke?

---

[2] Kit Fine makes a version of this point in Fine (1984).

5.3   Constructivist vs. Interpretationalist Options

5.3.1   The Constructivist Option

One option, suggested by taking talk about generating sets at face value, would be to say that sets are literally brought into being – perhaps by some act of social construction, like that which creates contracts and corporations.[3] For example, one might say that adopting an acceptable new axiom of set theory suffices to socially construct or extend the hierarchy of sets to a sufficient height to satisfy all of one's (now expanded) set-theoretic axioms.

One way of developing this social constructivist approach would involve biting the bullet and rejecting the idea just mentioned, that sets exist metaphysically necessarily. One might propose an error theory about why we falsely think the sets exist necessarily along the following lines. It sounds odd to deny that sets exist necessarily and timelessly because (as noted above) in all normal mathematical contexts apparent claims about set existence really express modal claims (as per the Parsonian paraphrase strategy). And the Potentialist paraphrase of the claim that some sets exist really is a timeless necessary truth.

Alternately, one might reconcile the that sets are socially constructed with the idea that all mathematical objects are metaphysically necessary and timeless, by drawing on some ideas from Cole (2013) and Searle (1995) about social construction and the possibility of decisions (about when a company came to exist, or when a player first qualified as on the injured list) taking effect retroactively.[4]

However, I take it that significant work would be needed to develop either of the above positions. So, it's not surprising that existing Parsonians tend to take a different approach.

5.3.2   Interpretational Possibility

Fine (2005, 2006) proposes a notion of interpretational possibility which has been taken up by the two most developed versions of Parsonian set theory in the current literature.

Fine introduces the notion of interpretational possibility by a kind of idealization on claims about how it is (physically or metaphysically) possible to reinterpret a given speaker. He suggests that certain acts of reinterpreting a speaker (e.g., by taking their quantifiers to range over an additional layer of sets) witness, but are not necessary for, the interpretational possibility of there being more sets. For example, Fine (2006) writes:

[I]t seems clear that there is a notion of [of possibility] such that the possible existence of a broader interpretation is ... sufficient to show that [a] given narrower interpretation is not

---

[3] See Cole (2013) for a proposal that mathematical objects are socially constructed in the same way as marriages and corporations. But note that Cole doesn't take mathematical objects to have the temporal features needed to drive the Potentialist story.

[4] Both have suggested that objects which are contingently socially constructed at a certain time (e.g., human rights constructed by a court) might nonetheless be necessary and exist eternally.

absolutely unrestricted. For suppose someone proposes an interpretation of the quantifier and I then attempt to do a "Russell" on him. Everyone can agree that if I succeed in coming up with a broader interpretation, then this shows the original interpretation not to have been absolutely unrestricted. Suppose now that no one in fact does do a Russell on him. Does that mean that his interpretation was unrestricted after all? Clearly not. All that matters is that the interpretation should be possible. But the relevant notion of possibility is then the one we were after; it bears directly on the issue of unrestricted quantification, without regard for the empirical vicissitudes of actual interpretation.

Fine contrasts the notion of interpretational possibility with "circumstantial" modalities like physical and metaphysical possibility. Contingent differences to what the world is actually like are supposed to make no difference to interpretational possibility. Fine writes, "Circumstance could have been different; Bush might never have been President; or many unborn children might have been born. But all such variation in the circumstances is irrelevant to what is or is not [interpretationally] possible" (Fine 2006). And many different things are interpretationally possible relative to the actual world (as, perhaps, witnessed by the fact that we could interpret someone as speaking with implicit quantifier restrictions and talking about more or fewer of the objects we are currently quantifying over). Interpretational possibilities are supposed to be (as Fine puts it) a kind of "possibilities for the actual world," rather than "possible alternatives to the actual world."

Accordingly, there is no conflict between saying it's metaphysically necessary that the hierarchy of sets stops at a certain height and that it's interpretationally possible for it to have a different height. And Interpretationalist Parsonians see no tension between understanding possible set formation in terms of interpretational possibility and accepting the intuition that sets exist necessarily.

Fine ultimately rejects understanding mathematics in terms of interpretational possibility, but (as noted above) both Linnebo and Studd invoke use his notion of interpretational possibility to develop their versions of Parsonian set-theoretic Potentialism.

5.4 Linnebo and Studd

5.4.1 Linnebo's Interpretational Possibility

Linnebo (2018a) develops a version of Parsonian Potentialist set theory which invokes the above notion of interpretational possibility and connects it very directly to Frege's notion of abstraction principles. He develops a version of Parsonian Potentialism within a larger account of how we can shift our language (to conceptualize the actual world in terms of more objects) by adopting abstraction principles. He suggests that some objects are "thin" (with respect to some other objects), in the sense that we can come to know things about the former thin objects by introducing abstraction principles that specify identity conditions for them by appeal to the objects they are thin with respect to. For example, in Frege's classic case, if you are already talking about lines, you can start talking in terms of the abstract objects we call "directions," by stipulating that two lines have the same direction iff they are parallel.

Accordingly, we can interpret talk of "forming" new objects as making claims about how one could (re)conceptualize the world as containing additional objects. Linnebo writes that he will take "modal operators $\Box$ and $\Diamond$ to describe how the interpretation of the language can be shifted – and the domain expanded – as a result of abstraction" (Linnebo 2018a). In particular, $\Diamond \phi$ is supposed to be true if you could make ϕ true via some well-ordered sequence of acts of reconceptualizing the world via adopting abstraction principles (whether or not it would be metaphysically possible for anyone to make such a sequence of abstractions).

Note that the adoption of such abstraction principles doesn't bring anything into being – whether it be a physical object or an abstract object. Rather it involves "reconceptualizing" the world. Also note that Linnebo only considers reconceptualizations which recognize more objects, not ones which remove objects we currently recognize. Since his notion of possibility only allows the world to grow, it doesn't satisfy S5 (unlike logical possibility)[5] and Linnebo accepts the converse Barcan-Marcus formula as true with regard to interpretational possibility.

Importantly, Linnebo appeals to a notion of **dynamic** abstraction, which lets one expand the application of some previously understood notion by adopting an abstraction principle. One can, in effect, introduce a predicate "Old()" that applies to all of the objects one is currently thinking in terms of and then adopt abstraction principles that say that for every plurality of old sets there's a "set" collecting exactly these objects. We might think of the above abstraction principle as saying, "I'll continue to refer to all these old objects and start accepting certain abstraction sentences implying there are new ones which relate to the old objects in a certain way." This has the important effect that repeatedly adopting (syntactically) the same abstraction principle can lead you to talk in terms of longer and longer hierarchies of sets.

Finally, Linnebo holds you can only start thinking in terms of a set if you are already (or simultaneously start) thinking in terms of its elements (paradigmatically a set is introduced by adopting abstraction principles that say that there's a set collecting every plurality of old sets). This gives us the dependence of sets on their elements referenced in Section 5.2.

Now, like Hellman, Linnebo shows that we can justify the use of the ZFC axioms from certain modal assumptions – in this case, assumptions about what's interpretationally possible. However, it seems to me that some of the assumptions used in this proof raise an important question about how Linnebo's notion of interpretational possibility is to be understood. In particular, Linnebo appeals to the maximality principle:

Maximality: At every stage, all the entities that can be introduced are in fact introduced.

---

[5] Speaking in terms of Kripke models, when it comes to interpretational possibility only worlds that preserve or add to the objects existing in a world w_0 are accessible from w_0.

But this principle seems very implausible on the intuitive understanding of interpretational possibility as possibility with respect to how "the interpretation of the language can be shifted – and the domain expanded – as a result of abstraction" evoked above. For surely, we don't introduce all possible abstraction principles at once!

Perhaps one can solve this problem by simply understanding Linnebo's notion of interpretational possibility more narrowly (as concerning how the world could be reconceptualized at some stage of a process that *did* simultaneously introduce all possible abstraction principles at each stage). However, such ad hoc restriction of the intuitive notion of what can be got to by introducing abstraction principles above can make the concept of interpretational possibility seem significantly less principled (and hence less attractive as a choice of theoretical primitive) than logical possibility.

5.4.2   Studd on Interpretational Possibility

Studd (2019) develops the notion of interpretational possibility in a way that (I'll suggest) raises a similar concern about whether interpretational possibility is an attractive choice of theoretical primitive.

Interestingly, Studd introduces his notion of interpretational possibility by contrasting it with logical possibility, saying that interpretational possibility is "importantly similar and importantly different to logical necessity. Like logical necessity, it concerns possible shifts in interpretation rather than circumstance. But unlike logical necessity, the shifts in interpretation that are admissible are more closely constrained: not every logically-possible interpretation need be counted admissible." He then says his notion of interpretational possibility is very similar to Linnebo's. However, where Linnebo talks about interpretational possibilities as corresponding to ways of reconceptualizing the world, Studd talks about "admissible interpretations" for a certain lexicon. And where Linnebo talks about abstraction principles, Studd talks about successful attempts to liberalize our language by, so to speak, expanding the domain of quantification and adding some of these new objects to the extension of certain terms like "set."

Studd writes, "The truth of ϕ depends on whether the proposition that would be expressed by ϕ under other admissible interpretations of the lexicon is true (in the actual world)" and says the following about admissible interpretations:

Admissible interpretations result from shifts of interpretation of the kind that a [quantifier] relativist may bring about in her attempt to expand the universe. Such interpretations come with a natural ordering: an admissible interpretation j is said to succeed another i iff j results from one or more relativist attempts to admissibly liberalize the interpretation i. In this case we also say that i precedes j.

Studd notes that his concept of "admissible interpretation" (and hence interpretational possibility) differs from the natural language notion of what we could get our words to mean.[6] Instead of appealing to some such natural language notion, he uses Kripke-like

---

[6] He writes, "In the case of the present version of English, for instance, there's nothing to stop us from attaching new meanings to terms like 'set' and 'element' that are wholly unconnected with their current meanings. The new interpretation could reinterpret these terms to be coextensive with 'sloth' and 'eats' (as

models and a principle that these models must satisfy certain monotonicity and stability requirements to (somewhat metaphorically[7]) convey his concept of admissible interpretations and liberalizations. This amounts to requiring that relevant meaning change events have *at least* the following features: they only introduce new objects to the domain of the quantifiers and the extension of "set" and "element" without stopping the quantifiers from ranging over anything you are currently quantifying over (Monotonicity) or changing how "set" and "element" apply to these current objects (Stability).[8]

However, these constraints cannot be *all* that's required of admissible interpretations. For example, expansions of a language which add a new object to the extension of "set" and not the extension of "element," despite that language already recognizing an empty set (so the extensionality axiom is violated), are presumably not admissible interpretations. And later, when justifying mathematicians' use of the ZFC axioms, Studd makes a plenitude assumption that amounts to saying that whenever you liberalize the meaning of set to "add" one set, you must thereby add (at least) a full layer of new sets.

Overall Studd says rather little about how he understands interpretational possibility, beyond the points summarized above and some further principles specific to set theory. It's not clear to me that any single unified intuitive notion implies all the constraints on interpretational possibility Studd asserts (or whether Studd even claims to have latched on to such a notion). Thus, I think Studd's notion of interpretational possibility can, like Linnebo's, seems unprincipled and ad hoc in a way that makes it

the latter terms are presently interpreted). The resulting interpretation is clearly available to us but inadmissible because it fails to meet the Stability constraint. All the same, since this sort of reinterpretation is clearly orthogonal to issues concerning absolute generality, nothing is lost by taking the interpretational modal operators to only generalize over admissible interpretations."

[7] There couldn't actually be models witnessing all interpretational possibilities, without all the problems of arbitrary stopping points etc. re-arising.

[8] More specifically, Studd explains his notion of interpretational possibility metaphorically by providing something like a Kripke a model for his modal notion (with objects called indexes corresponding to specific interpretational possibilities). After conjuring the image of stages in a growing hierarchy of sets with indexes corresponding to particular stages of growth, Studd writes:

Less metaphorically, we can helpfully think of the indices [of these Kripke models] as admissible interpretations of the sort that the modality is intended to generalize over, with Monotonicity and Stability serving to constrain the sorts of interpretation the modality generalizes about.

In this model we have a set i of indexes for admissible interpretations i_1 i_2 etc. for "S" and "E" (for "set" and "element of"):

a model-theoretic (or MT-) interpretation of the non-modal language L_{SU} is a set-structure $\langle M, S, E \rangle$ that supplies a non-empty set M as the universe of discourse, and extensions S and E based on M for the language's two non-logical predicates, the set and element–set predicates (β and ϵ).

... an MT-hierarchy is an indexed-set of triples $\{\langle M_i, S_i, E_i \rangle\rangle : i \in I\}$ each member of which is either an MT-interpretation or the empty interpretation (with M_i non-empty for some $i \in I$), and which meets the ... three conditions [serial well order, monotonicity and stability]

Monotonicity. "Whenever i and j are indices in I with $i <_I j$, M_i is a subuniverse of M_j (i.e. $M_i \subseteq M_j$)."
Studd also has a requirement of stability which requires that any pair of admissible interpretations agree on the relation "element-of" on objects both interpretations recognize as sets.

Stablity. "Whenever i and j are indices in I, the extensions S_i and S_j and the extensions E_i and E_j agree on their common domain $M_i \cap M_j$."

unappealing as a choice for mathematical foundations (even if it's no problem for the project of defending quantifier relativism which most interests Studd (2019)).

5.5   Which Framework to Use?

With this picture of the most developed versions of Parsonian Potentialist set theory in mind, I will now attempt to motivate my choice to work in the Putnamian framework instead (at least for temporary practical purposes).

In this section I'll argue that Parsonians face some extra pressure (beyond the general reasons for endorsing a notion of logical possibility discussed in Section 4.1) to accept the basic logical machinery needed to state Putnamian paraphrases (and perhaps to agree with Putnamians about how these concepts apply). Then I'll explain (rather abstractly) why working in a Putnamian framework will be convenient for my justificatory project.

5.5.1   Acceptability of Logical Possibility to Parsonians

First note that Parsonians Linnebo and Studd do seem to accept the meaningfulness of logical possibility and seemingly agree that (at least some versions of) Putnamian paraphrases have the correct truth-values. As we saw above, Studd introduces his notion of interpretational possibility by appeal to logical possibility, and he even seems to (in some sense) endorse the adequacy of Hellman's Putnamian Potentialism.[9] And Linnebo's criticisms of Hellman-style Putnamian Potentialism in his head-to-head comparison of Putnamian and Parsonian Potentialism in Linnebo (2018b) are strikingly moderate and don't center on raising doubts about the intelligibility of Hellman's proposal.[10]

---

[9] In a footnote to his chapter on Potentialist set theory, Studd writes the following about Hellman's modal Structuralist approach (and gives no later criticism of that view):

> An alternative is for the relativist to adopt modal-structuralism in the style of Hellman (1989). This permits her to interpret □ simply as logical necessity. On this view, there is no need for admissible interpretations to satisfy the Stability constraint on the interpretation of the non-logical vocabulary set out below. This is because modal structuralism takes set-theoretic statements to be elliptical for statements in a higher-order modal language, which eliminates occurrences of the set and element–set predicate. See also Hellman ... who applies modal-structuralism to offer a potentialist Zermellian response to the set-theoretic paradoxes.

[10] After raising the metaphysical shyness and compossibility worry we discussed in Section 3.3.3, Linnebo notes:

> Let me be very clear about my complaints in this section. I am not asserting that metaphysically shy objects are in fact possible or that there might not be some clever way to ... circumvent the problems generated by the phenomenon of incompossibles. My point is only that the extra freedom of Putnam's approach, which initially seemed purely advantageous, has the unintended side effect of incurring potentially problematic metaphysical commitments, which are avoided on the Parsons approach.

> Linnebo's other points against versions of the Putnamian approach in that paper merely involve correctly pointing out version of some points already discussed above: that it's hard to make sense of Putnam's dual perspective (mathematics being equally well understandable in modal and ontological terms), and that Hellman's requirement that initial segments be models of ZFC_2 seems unmotivated and troublesome.

I think this apparent willingness to accept the meaningfulness of a notion of logical possibility (and something like the intuitions about it the Putnamian Potentialist needs to appeal to) is no accident, but rather flows from something basic about Putnamian paraphrases.

For, note that both Studd and Linnebo take for granted the notion of well-founded sequences of reconceptualization/liberalization events in developing their concepts of interpretational possibility and Potentialist set theory. Arguably any modal notion that could do the work the Parsonian needs must appeal to a very idealized notion of how some objects could be well ordered by a relation. But it's hard to see how one could understand the relevant modal notion of a possible arbitrary well-ordered sequence of language changes, without a background notion of something like logical possibility.[11]

Pressure for the Parsonian to accept a notion of how it would be possible (in some sense that isn't hostage to facts about metaphysical possibility) to have a sequence of set-formation events satisfying the well-ordering axioms is clearest if we understand the Parsonian to allow only adding a single layer of sets at each stage. But, even if we allow arbitrarily many sets to be introduced at any stage, the Parsonian would be hard pressed to try and insist that it's enough to consider only a well-ordered sequence of reconceptualization events of some limited height.[12]

Thus, one might argue that the Parsonian already needs to understand all the notions needed for Putnamian Potentialism. Indeed, one might argue that the Parsonian is already appealing to a Putnamian Potentialist picture of the *ordinals* to motivate their story. For, the Parsonian must already take there to be a fact of the matter about what well-ordered sequences are possible in some sense that's obviously meant to be free of any purely physical or even metaphysical limitations. If they are going to accept the meaningfulness of asking if there is a well-ordered sequence with a certain property, it would be unattractive to suggest that such talk of coherence or logical possibility is only meaningful for well-orderings and nothing else.[13]

Finally, I've argued that Linnebo's shyness-based criticism of Putnam's account can be avoided by reformulating Putnamian proposals using the conditional logical possibility operator as per Chapter 3. Some Nominalists might worry about implicit commitment to abstract objects – but Parsonian Potentialists who embrace the existence (or at least possibility) of sets will not have that doubt.

[11] Interpretationalist Parsonians could, of course, reply that they use the notion of logical possibility to explain interpretational possibility and then kick away the ladder, just as I've suggested one could use appeal to Actualist set theory, isomorphisms between possible worlds or Shapiro's structures to point at the notion of conditional logical possibility, without endorsing any of these views. But they don't seem inclined to do this, and this move is somewhat awkward for the reasons noted below.

[12] For example, suppose that all relevant height increases could be performed by repeating a single set-generating ceremony $< \alpha$ many times. In Actualist terms this would amount to assuming that the whole hierarchy has cofinality α (something widely regarded as implausible by mathematicians). For, in Actualist terms it would imply that some second-order function f, and perhaps even some first-order definable relation ϕ, could map α to all the ordinals, contrary to the (the spirit of, and second-order formulations of) axiom of replacement! And a similar conclusion follows if the menu of different abstraction techniques one can in principle apply is small relative to α.

[13] Note that the second-order/plural quantification vocabulary typically used to formulate the claim that a sequence of growth events is well ordered also lets you pin down intended models of the iterative hierarchy up to width. So they need to accept not just the logical possibility of first-order facts but (something like) logical possibility claims about a hierarchy of growth/reconceptualization events satisfying the non-first-order least element condition of well foundedness.

Accordingly, I take it that Parsonian Potentialists generally do and plausibly should accept the meaningfulness of logical possibility (together with other tools needed to develop Putnamian set theory).

5.5.2   Putnamian Potentialism and Logical Possibility

This is a happy result, because it means that if we succeed in justifying the Putnamian Potentialist version of the axiom of Replacement from principles that seem clearly true, Parsonians can plausibly use this result to (partially) justify *their* version of the axiom of Replacement by inferring it from the Putnamian version of Replacement.

Working in a Putnamian framework (of the kind advocated in previous chapters) – at least temporarily in the sense above – turns out to be quite convenient, for my proposed justification of the axiom of Replacement leverages intuitions about the logical possibility of structures other than initial segments of the set-theoretic hierarchy in a way that seems difficult, if not impossible, to reproduce using a more narrowly tailored notion of possibility like that advocated by Parsonians.

Specifically, we will often justify the Potentialist translations of set-theoretic claims (claims about how iterative hierarchies can be extended by other iterative hierarchies) by first proving things about how any such hierarchy could be extended by certain larger structures that *aren't* iterative hierarchies. But it's difficult to see how to reconstruct such reasoning about extensibility working purely within in an Interpretationalist Parsonian framework, where growth events seemingly only add objects falling under some currently understood indefinitely extensible concept[14].

Negatively, however, the fact that Parsonians seem to endorse the basic machinery needed for Putnamian set theory raises an ideological Occam's razor worry for the Parsonian. If we already need to accept a notion of logical possibility sufficient to formulate Potentialist set theory, and use reasoning about logical possibility as a first

---

[14] For example, the Parsonian might try to mirror the reasoning above by considering the interpretational possibility of a hierarchy of sets existing alongside other (non-set) objects. But note, it seems that there might not be any concepts in our current language whose rich meaning (in the sense specified in Section 5.6.2) allows them to form the kind of structure we want to consider our iterative hierarchy embedded in for the kind of roof considered above. It seems implausible that, for every describable way it would be logically possible for some relations $R_1, \ldots, R_n$ to pick out a larger structure it's useful to consider a hierarchy of sets being embedded within, it's interpretationally possible for some relations $R'_1, \ldots, R'_n$ to apply in exactly that way. Recall that Studd and Linnebo take there to be various important facts about the meaning of set and element, which ensure that, e.g., you can't think in terms of multiple sets that have exactly the same elements. And presumably the same applies to other current English language concepts as well. Perhaps we could get around this problem by considering interpretational possibilities corresponding to language changes that add *new* atomic predicates and relations to our language.

But this approach is difficult to develop in Linnebo or Studd's system. For Studd's talk about interpretational possibility reflects admissible interpretations of "the" lexicon (rather than considering a lexicon that could be arbitrarily extended). And, while Linnebo's system seems to be more open to introducing new concepts, he says interpretational possibilities correspond to the way we could start talking in terms of new objects by adopting abstraction principles, and it's not clear that every larger structure which it is useful to reason about initial segments being embedded in (for the purposes of non-elementary proofs as above) can be introduced by stipulations which take the form of abstraction principles.

step in justifying certain axioms of Potentialist set theory, why bother cumbering our ideology with the notion of interpretational possibility? Why not just use the same notion to formulate Potentialist set theory? I will consider this concern as part of the question of which view to ultimately prefer in the next section.

5.6   Which Framework to (Ultimately) Choose?

Now let's turn to the question of whether Putnamian or Parsonian Potentialism is ultimately to be preferred. Admittedly, this question is a little ambiguous; you might ask, "which version of Potentialism should we prefer *for what purposes*?" For example, we might ask (in a Sideran pro-metaphysics spirit), which formalization of set theory best reveals the facts about fundamental ontology and ideology that ground the truth of set-theoretic claims. Alternately one might ask which theory provides the best Carnapian explication of Potentialist set theory. However, my remarks below will motivate favoring Putnamian Potentialist set theory in both of these ways, so I won't stress the distinction here.[15]

5.6.1   Ideological Parsimony and Conservatism

One kind of motivation for favoring Putnamian set theory draws on considerations of ideological parsimony. In Sections 4.1 and 4.5 we saw some reasons to think that even Parsonians should accept the Putnamians' notion of logical possibility. Accordingly, as noted above, one might think Putnamian Potentialism should be favored on grounds of ideological parsimony (let us not multiply primitive modal notions beyond necessity) and avoiding revisionary and controversial commitments about other areas of philosophy where possible. If Parsonians must accept Putnamian primitives but not vice versa then parsimony surely favors the Putnamian approach.

Admittedly Parsonians could block this argument if they could show that Putnamians also can't do without Parsonian primitives. Interpretationalist Parsonians might argue that we independently need the notion of interpretational possibility to make sense of neo-Carnapian language change. However, in Chapter 16, I'll argue that this is not the case. We can develop a neo-Carnapian philosophy of language sufficient to do the (legitimate) work of neo-Carnapian philosophy of language equally well or better using the conditional logical possibility operator. Notably, I'll suggest that doing this legitimate work doesn't require stating claims about the possibility or impossibility of "absolute generality"/quantifying over anything in some non-trivial sense (the main thing, outside set theory, Linnebo and Studd use the interpretational possibility operator to do).

---

[15] Note that some traditional reasons for favoring Platonistic views over modal perspectives on mathematics don't bear on our choice here. Neither Parsonian nor Putnamian Potentialists take set theorists' apparent quantification at face value, and both introduce new modal notions that go beyond FOL in analyzing set-theoretic claims.

In addition to adding a new modal operator, Interpretationalist Parsonianism also requires us to make some prima facie unintuitive changes to philosophy of language.[16] Linnebo himself the following (italics mine).

Suppose we have formulated a perfectly precise notion of a star. For any object whatsoever, this notion enables a definitive verdict as to whether or not the object is a star. When this precise intension is applied to the world, reality answers with a determinate extension, namely the plurality of objects that satisfy the intension. And there is nothing unusual about stars in this regard. *In most ordinary empirical cases, a precise intension determines an extension when applied to the world. But in mathematical cases, and other cases involving abstraction, this is no longer so. Here a precise intension often fails to determine an extension.* (Linnebo 2018a: 69)

Thus, overall, one might argue that the main motivation for accepting interpretational possibility (and the claim that precise intensions don't determine precise extensions above) is to account for set-theoretic paradoxes while avoiding arbitrariness intuitions.[17] But, if a Putnamian approach can do the same work without requiring us to add to our fundamental ideology or revise general philosophy of language, considerations of parsimony favor the Putnamian framework.

5.6.2   Unappealing Choice of Conceptual Primitives

Another argument against (Interpretationalist) Parsonianism questions the attractiveness of interpretational possibility as a conceptual primitive (as compared to logical possibility). In this section I'll note some ways that interpretational possibility can seem much less principled and clearly/concretely understood than the notion of logical possibility, and hence like a less attractive choice of a conceptual primitive for philosophical analysis.

In Sections 5.4.1 and 5.4.2 we already saw some reasons for concern that Linnebo and Studd's notions of interpretational possibility must be arbitrarily restricted/non-joint-carving in ways that makes them a bad choice for a conceptual primitive (if they are to satisfy the various assumptions Linnebo and Studd use to vindicate use of the ZFC axioms). Additionally, Linnebo himself notes a way that interpretational possibility facts reflect arbitrary conventions with respect to "Julius Caesar problems" about when the objects falling under the concept introduced or liberalized by adopting some

---

[16] Going further, Linnebo's embrace of something like Dummett's indefinite extensibility in the quote above may raise prima facie philosophical puzzles, which could easily be avoided by favoring Putnamian Potentialism. For it seems to me that the passage above suggests the following picture. There aren't just different *equally legitimate* ways of "talking in terms of more sets" (for in this case our concept of a hierarchy would merely not be "precise") but rather we have a precise concept with a kind of inadequacy or internal tension, whereby every language including the concept set is held to be in some way *leaving some things out* so that, e.g., languages that talk in terms of more sets are less inadequate than languages that talk in terms of fewer sets. But such ideas about reality forever transcending language and thought can seem prima facie problematic and, and hence desirable to avoid where possible.

[17] Certainly the case of set theory is the main motivation cited for Dummett's project in Dummett (1993) and Studd's project of defending quantifier relativism in Studd (2019).

abstraction principle (e.g., the number 1) are identical to objects one was previously talking in terms of (e.g., Julius Caesar).[18]

Interpretational possibility facts also reflect seemingly highly indeterminate and/or disputed facts about what I'll call "rich meanings," in a way that can make the interpretational possibility operator seem like an unattractive choice of primitive. The Interpretationalist Parsonian needs to distinguish between acts of neo-Carnapian language change that count as talking about more *sets* vs. beginning to use the word "set" to express some other concept. For example, they can't allow that it's interpretationally possible that "there are two sets with exactly the same elements," although obviously one could change the meanings of English words, so the corresponding sentence expressed a truth. And presumably the current meaning of the word "set" (perhaps together with background linguistic conventions and precedents) is what does this work.[19]

Accordingly, Interpretationalist Parsonians seem to be committed to our words like set having "rich meanings," which specify what's needed to preserve the meaning of a word (to continue talking about *sets*) under some neo-Carnapian language change that gets us to start talking in terms of new objects (e.g., by introducing an abstraction principle).

Now I admit that we have *some* shared and correlated intuitions about how the meaning of the terms "set" and "element" could be preserved neo-Carnapian language change. However, it seems to me that such agreement is limited and vexed in the same ways as agreement on the right way to expand the meaning of your terms for the purposes of engaging with a metaphor.

For example, I take it that most people might agree that saying the leader of a country is its "head" is a reasonable way to preserve/honor the current/literal meaning of the term "head" in a metaphorical context which invites us to apply

---

[18] He writes, " . . . when we develop our linguistic practices, we have some degree of choice about whether or not to allow categories to overlap. To handle mixed identity statements, we often need conceptual decisions, not just factual discoveries . . . . When our ancestors first confronted Caesar-style questions [i.e., questions like whether Julius Caesar is identical to the number one], they had a choice which way to go; and this choice played a role in shaping the concepts that they thereby forged. Today we find ourselves in a different situation, since many choices are already implicit in the linguistic practices that we have inherited. Of course, insofar as we are willing to revise these practices, we still have the same choice as our ancestors had. But we face an important additional question not encountered by our pioneering ancestors, namely what conceptual decisions are implicit in our inherited linguistic practices. I shall argue that these practices have by and large legislated against the overlap of categories. But exceptions are certainly possible and very likely even actual" (Linnebo 2018a).

[19] An analogous argument can be made even if we assume that interpretational possibility must satisfy Studd's Maximality and Stability assumptions. For (if you accept the neo-Carnapian view at all) it's intuitively possible to change your language/thought so as to add a new object to the extension of "set" and not the extension of "element" (so, given that we're already talking in terms of empty set, the extensionality axiom begins to express a falsehood). But the Interpretationalist can't allow that it's interpretationally possible for there to be two *sets* with exactly the same elements (if the interpretational Parsonian allowed this, their paraphrases of the axiom of extensionality would come out false). Thus, there must be some reason that changing your language use in the way indicated above only says something about how one could change the meaning of the word "set" and not about what *sets* it would be interpretationally possible for there to be. And presumably the current meaning of the word "set" is what does this work.

human anatomical language to parts or aspects of a country. But this limited agreement doesn't provide (or evidence) shared understanding of a sufficiently precise and concretely grasped notion of metaphorical truth (or possibility) to make the latter concept an attractive choice of primitive when logically regimenting mathematics.

Our intuitions about rich meanings, even in the Interpretationalist's key case of the concepts *set* and *element* can seem similarly limited. I take it that most people would agree that if we think in terms of more sets, it's natural to suppose these sets would still have to satisfy extensionality. But suppose I am currently thinking in terms of certain hierarchy of sets V_α. If I were to adopt an abstraction principle which adds an extra layer of "sets," would this really be a way of thinking in terms of more *sets* (on my current meaning of the term), as the Parsonian account needs? It seems equally or more natural to say that, after making this switch, only the "sets" up to V_α are sets on my current sense of the term, and when expanding my quantifiers in the way suggested I have instead got the word "set" to express a new concept like "class." Note that, e.g., the new sets thus introduced won't satisfy the pairing axiom.

5.6.2.1 Object Identity under Neo-Carnapian Language Change

A similar point about interpretational possibility facts being seemingly controversial and/or indeterminate *may* arise in connection to judgements about *object* preservation under neo-Carnapian language change. Linnebo and Studd can seem to endorse generally determinate facts about when different sequences of abstraction principles wind up introducing the same object.[20]

Yet it can seem implausible that there are, in general, such determinate facts about when adopting one sequence of abstraction principles introduces the same entity you could have introduced by some other sequence of abstraction principles.

For example, are the objects you would have introduced by introducing the concept of "Turing degrees" by abstraction over computations formalized using Turing machines *literally the same entities as* those introduced via abstraction over computations formalized using general recursive functions? And suppose someone introduces a concept of "Turing degrees" via abstraction principles involving an unspecified notion of one set of numbers being "computable from" another which is equally anchored to both definitions of computability. Is this a way of introducing *literally the same entities* you could have introduced by introducing "Turing degrees" in one of the two ways mentioned above (or merely some isomorphically structured mathematical objects)?

We do, sometimes, say that people whose mathematical definitions differ slightly from ours can "know things about" structures like the natural numbers or Turing

---

[20] For example, in motivating a certain convergence assumption about interpretational possibility Linnebo (2018a: 206–207) writes:

This principle ensures that, whenever we have a choice about which entities to introduce, the order in which we choose to proceed is irrelevant. Whichever entity we choose to introduce first, the others can always be introduced later. Unless $\leq$ was convergent, our choice about whether to extend the ontology of w_0 to that of w_1 or that of w_2 would have an enduring effect.

degrees. But arguably what's required for such claims to be true is highly indeterminate and/or context dependent. For example, in most situations it seems reasonable to describe people who know that some claim ϕ holds for any of the Turing-degree-like mathematical structures introduced by any of the acts of abstraction above as knowing something "about the Turing degrees." However, in contexts in which the equivalence of Turing machines and general recursive functions can't be assumed as background knowledge, we may draw finer distinctions.[21]

Note that many philosophers like McGee (1997) find it positively attractive to say that reference for abstract terms like natural number (a paradigmatic case of objects Linnebo and others want to say could be introduced via abstraction principles) is only determinate up to isomorphism. Thus, we appear to have another dimension along which facts about interpretational possibility must be controversial, arbitrary and/or indeterminate.

5.6.2.2   Veil-lifting Picture

Admittedly, there is a certain picture which might motivate thinking there are principled determinate answers to the questions about rich meaning and the identity of objects introduced by abstraction principles above. However, this picture has other, very unattractive features. I will conclude this subsection by discussing it, although I don't mean to claim Linnebo or Studd would endorse it.[22] My point is only that, unless we take everyone to be unveiling portions of some shared total world (in the sense sketched below), it's unclear what would explain there always being definite facts about which Carnapian language changes preserve the meaning of predicates or wind up introducing the same object.

Veil-lifting Picture: There's a shared total world containing all the different kinds of objects anyone could ever talk in terms of. The meaning of each atomic predicate or relation determines its extension within this total universe. Acts of neo-Carnapian language change get you to talk in terms of more objects by lifting parts of the veil covering a plentiful universe, expanding the domain of objects your quantifiers ranger over to include more things. You can't ever reach a point at which you couldn't expand your quantifiers further. But this is just because (for some reason) no series of linguistic acts could lift the veil completely, so that your quantifiers would range over absolutely all objects that one could, in principle, talk in terms of.

If we take this picture of neo-Carnapian language change as quantifier restriction lifting seriously, we automatically get the required rich meanings (determinate facts about how each property we currently talk in terms of will apply under quantifier meaning shift) and determinate facts about which acts of re-conceptualization by adopting abstraction principles would introduce the same objects. For example, there will, prima facie, be determinate facts about whether people who "push back the veil"

---

[21] Also it's appealing to say that a pair of people who identify the numbers with different ω-sequences of sets (*a la* Benaceraff's famous paper (Benacerraf 1973)) both still "know things about the numbers." But one *can't* say this is true in virtue of them literally talking about the same objects/numbers, since number 3 can't be identical to two different sets.

[22] Both certainly say that they mean interpretational possibility to reflect an expansionary modality rather than mere removing of quantifier restrictions from some fixed universe.

by introducing "Turing degrees" via abstraction principles involving general recursive functions vs. turing machines are unveiling the same objects.

However, endorsing general determinate facts of this kind can seem unattractive. For note that this veil-lifting picture adds a presumption of determinacy, which goes beyond traditional metaphors for neo-Carnapian language change, on which "different languages carve up the world into objects in different ways."[23] In Chapter 16 I will propose a more concrete approach to neo-Carnapian language change which also avoids this presumption of determinacy.

One might even fear that adopting this veil-lifting undermines the motivations for Potentialism discussed in Section 1.2. For on the veil-lifting picture it appears there's some shared ineffable domain of all objects which all neo-Carnapian language shifts correspond to quantifying over portions of; we just can't succeed in forcing our quantifiers to be interpreted as ranging over all this structure. But the intuition driving Potentialism in Section 1.2 was that for *any* actual plurality of objects there could be a larger one, not just that for any plurality of objects *we can get our quantifiers to range over* there could be a larger one. So, I think it's desirable to avoid the above veil-lifting picture if we can.

So, to summarize, facts about interpretational possibility can seem unprincipled (not joint-carving), non-fundamental, disputed and/or frequently indeterminate in ways that make interpretational possibility an unattractive primitive for mathematical or philosophical analysis.

5.6.3   Double Duty

5.6.3.1   The Problem

Finally, a third challenge for Parsonians concerns the double duty set talk is supposed to play in Parsonian theories.

Recall that the Putnamian Potentialist never employs a predicate "set" in her logical regimentations of set-theoretic talk. Thus, she could deny that there is a property of being a set. Or if she does accept such a notion, she can say that it has an empty extension, for the same reason "phlogiston" does (sets are objects hypothesized by the wrong account of what set theory teaches us about).

In contrast, the Parsonian takes ordinary mathematical usage to have given "set" a definite meaning – enough that there are definite (non-trivial) facts about how tall a hierarchy of sets actually exists/mathematicians are currently thinking in terms of. But it can seem puzzling how mathematicians' set-theoretic talk can do this while simultaneously being best understood in a Potentialist fashion. How can such talk

---

[23] If I carve up some dough one way, but I could have carved it up another way, we don't seem forced to accept determinate *de re* facts about whether a certain cookie brought into being by one carving could instead have been brought into being by another carving and baking sequence. A metaphysician like Kripke *might* choose to directly endorse such facts. But they might equally well analyze such facts away in terms of claims about contextually relevant counterparts, like Lewis. Or they might choose to reject such questions as meaningless (as Quine does in rejecting quantifying-in).

determine what height Actualist hierarchy of sets they are thinking in terms of (in any principled fashion)?

For Interpretationalist Parsonians, the challenge looks like this. In what sense can someone said to be "thinking in terms of" *any* hierarchy of Actualist sets with height α, if their set talk should always be interpreted potentialistically? How many sets are Linnebo and Studd currently thinking in terms of? And what principled grounds are there for the answer to this question?

To me it seems like the obvious thing to say is that we (Potentialist philosophers and mathematicians speaking in normal contexts where they should be understood potentialistically) aren't currently thinking in terms of any sets – that the interpretational possibility of some sets existing reflects what's possible via a quite different linguistic practice where someone would take the sets to stop at a definite point. But Linnebo and Studd don't seem to say this.

To press this worry (and further clarify Studd's notion of interpretational possibility), I will now discuss Studd's story about our set-theoretic practice might unknowingly get our quantifiers to range over more and more sets. It seems to me that if this story worked it might attractively answer the challenge above: providing a principled account of how many sets mathematicians are currently thinking in terms of. However, I will argue that it does not work, so the double duty problem remains.

5.6.3.2 Studd on Expansion and Actual Set Theory

Studd (2019: Chapter 8) sketches a story about how people with a set-theoretic practice like ours could unknowingly change their quantifier meanings and come to talk in terms of a progressively larger Actualist hierarchy of sets.

Studd first considers a situation where people *knowingly* start talking and thinking in terms of extra sets. Imagine that some people start out speaking a language Q. Then they decide to split off from the main body of Q speakers and develop a new language E, which "talks in terms of" extra sets.

To do this they adopt certain principles, most importantly the inference schemas below for reasoning from claims in the old language Q (indicated below by putting "Q:" in front of them) to claims in the new language E (indicated below by putting "E:" in front of them), and vice versa:

$$(U_E - \cdot)$$

$$Q : \text{things}(vv) \Rightarrow E : \text{thing}(vv)$$

$$Q : \text{things}(vv), Q : v \prec vv \Rightarrow E : v \in vv$$

$$Q : \text{things}(vv), E : v \in vv \Rightarrow Q : v \prec vv$$

Intuitively these schemas embody the idea that each plurality vv of objects quantified over in the old language Q is supposed to form a set in the new language.[24] Much might be said about such exotic principles. Note, for example, that vv is a plural variable. So, we aren't reasoning from *sentences* in one language we speak to another,

---

[24] See p. 235.

but supposing that we can (so to speak) reach out and catch the reference of a free variable in some formula in one language, by a formula in another language. But I take the general picture of accepting such inferences forcing a charitable interpreter to interpret the quantifiers in your new language E as ranging over strictly more objects than they did in your original Q to be clear. And I won't object to any of these details here. My objections concern the next part of the story.

With this background in place, Studd then considers how we can charitably interpret speakers who accept something like the inference rules above but have subtly incoherent beliefs:[25]

$$things(vv) \Rightarrow thing(vv)$$

$$things(vv), v \prec vv \Rightarrow v \in vv$$

$$things(vv), v \in vv \Rightarrow v \prec vv$$

The above inference principles let you infer that, for any plurality of things vv, there's a set vv whose elements are exactly the objects v in this plurality vv (written $v \prec vv$). Thus, accepting it (together with normal plural comprehension principles saying that for any ϕ there's a plurality vv of the objects such that ϕv) lets you derive the existence of the Russell set and hence contradiction.

Studd argues that these speakers could undergo a kind of unwitting quantifier meaning change, for the following reason. In general, a charitable interpreter can try to accommodate a speaker's reasoning by changing the domain of objects they take the speaker to quantify over[26] and the language they take them to be speaking. In this case, Studd suggests, charitable interpretation might take the speaker to be going through something analogous to the language switch from Q to E envisaged above. And if meaning reflects charitable interpretation, then we can have a kind of unwitting quantifier meaning expansion in this way.

This is, I take it, Studd's proposal for how it could be true that (unbeknownst to us) our current quantifiers range over some steadily growing range of sets. He puts it forwards as the "basis for an idealized account of universe expansion applicable to the ordinary English speaker." I have the following concerns.

First, surely we don't actually, after the discovery of Russell's paradox, have the disposition to infer that arbitrary pluralities form a set?

Second, and perhaps most importantly as regards answering the double duty problem, Studd's story doesn't suggest any principled answer to when and how quickly speakers are supposed to go through language change events he proposes. How often would the charitable interpreter say that someone with the inference dispositions above has switched languages? (Every 5 minutes? Every 10 minutes?) If I lie around, having the inconsistent inference dispositions Studd mentions and not thinking about set theory for an hour, how many times should the charitable interpreter take my language to have

---

[25] See p. 239.

[26] Studd gives this example, "I utter '52% of people voted for Brexit' and we immediately limit the domain to exclude those who didn't turn out or were ineligible to vote."

changed during that time? Insofar as standing dispositions to make inferences (or regard failure to make inferences as irrational) drive the above charitable interpretation, it is hard to see how one could give any non-arbitrary answer to the above question.

Third, it's not clear whether *the balance* of charitable interpretation favors Studd's strategy, once we fill in relevant speakers' *other* inference dispositions in a realistic way. For one thing, people are disposed to interpret things they wrote yesterday homophonically and assume that the truth-value of sentences depending on the height of the sets doesn't change from day to day. But Studd's favored charitable interpretation would make this trans-language inference schema fail.

For another thing, it seems to me there's a dilemma about what different stages of growth in the hierarchy of sets are supposed to look like. If the hierarchy of sets grows one layer at a time, then it looks like reinterpreting someone as talking about a larger hierarchy will sometimes be very uncharitable. For example, doesn't going from interpreting someone as talking about V_ω to $V_{\omega+1}$ make various things they believe like the pairing axiom come out wrong?

But if we avoid this problem by saying that each reinterpretation of set-theoretic talk must interpret people as quantifying over a domain of objects satisfying something like ZFC_2, we will "ascend in big leaps" like Hellman rather than in single steps as Linnebo and I prefer, and face the inconveniences discussed in Section 3.2.1.

Also, at the risk of sounding crude, why isn't Putnamian Potentialism (which, as we saw, Studd acknowledges the acceptability of) a more charitable interpretation than any of these? Why isn't the Parsonian interpretation itself a better interpretation of the speakers accepting incoherent inference procedures Studd imagines?

So, overall, I don't how to get any clear attractive answer to the question "how many Actualist sets am I currently thinking in terms of?" from Studd's account.

5.7   Conclusion

In this chapter I've discussed the differences between Putnamian and Parsonian approaches to Potentialism and reviewed some major forms of Parsonian Potentialism.

I've then tried to justify my use of a Putnamian framework to Parsonian readers who may find it unfamiliar. I've argued that Parsonians can and should accept the meaningfulness of basic concepts like logical possibility and can likely use my Putnamian version of Replacement to further justify their version of Replacement (at least to some extent). I've also (loosely) indicated the reasons why working in the Putnamian system will be practically useful.

More tentatively, I've argued that we should favor Putnamian over Parsonian approaches to Potentialist set theory on approximately the following grounds.

First, Putnamian Potentialism can be developed using a joint-carving notion of logical possibility which everyone has reason to accept (and Putnamians have extra reason to accept). Thus (Interpretationalist) Parsonian set theory can seem unparsimonious and needlessly revisionary, insofar as it requires adding an interpretational possibility operator to our ideology and make certain otherwise unneeded revisions to our philosophy of language (denying that "precise intensions always determine precise extensions").

Second, the notion of interpretational possibility can seem like an unattractive choice of theoretical primitive. For facts about interpretational possibility generally would seem to be frequently indeterminate, highly disputed and/or unprincipled facts. And Linnebo and Studd's particular versions of this concept can seem ad hoc restricted to allow assumptions needed to justify set theory. Thus, one might favor Parsonian Potentialism on grounds of conceptual parsimony. There is also a worry that the Putnamian Potentialist needs to – in effect – invoke a Putnamian approach to the ordinals (a notion of arbitrary sequences of reconceptualization events satisfying the axioms for being a well ordering), in which case adding a philosophically different Parsonian approach to the sets seems particularly unmotivated.

Third, Parsonians faced a "double duty" problem which Parsonian views avoid. For example, Interpretationalist Parsonians face awkwardness about how to answer the question "how many sets are you currently thinking in terms of?" that is not answered by Studd's picture of unintentional language change.

In closing I will mention three, weaker, motivations for favoring the Putnamian approach. First, Putnamian paraphrases promise to make the intuitively close relationship between math and logic explicit.

Second, the *practical convenience* of working in a Putnamian framework (and cashing out set theory in terms of logical possibility) discussed in Section 5.5.2, might be taken as evidence for the philosophical correctness of this approach. Going Putnamian promises to let us rationally reconstruct the justification for our set-theoretic beliefs from premises that seem clearly true more directly, using fewer primitives.

Third, although I personally think we should accept a broadly neo-Carnapian philosophy of language, it's worth noting that Putnamian Potentialist set theory doesn't require us to accept this controversial philosophical thesis while (Interpretationalist) Parsonianism does.[27] Thus, philosophers who reject such neo-Carnapian philosophy of language will certainly favor the Putnamian approach over the Interpretationalist Parsonian one. And perhaps the same goes for philosophers who would prefer to leave few "hostages to fortune" and avoid entangling the philosophy of set theory with unrelated philosophical controversies.

---

[27] The interpretational Parsonian can't mean interpretational possibility in the familiar Tarskian sense where all interpretations choose their domains from among some fixed universe of objects, otherwise we will have a maximum size which all interpretations of the sets have to be found within. On such a view, Actualists' apparent commitment to arbitrary stopping points, which Potentialism promised to let us avoid, seems to get dragged back in.

Part II

In Part I we discussed some existing Potentialist and Actualist approaches to the foundations of set theory and noted some problems for them.

In Part II of this book, I will develop my particular version of (Putnamian) Potentialist set theory using conditional logical possibility and argue that it lets us avoid many of the problems discussed above. As we have seen, Potentialist paraphrases of set theory make claims about how it would be (in some sense) possible to extend an initial segment of the hierarchy of sets.

First I will flesh out the informal summary given in Chapter 4 of how conditional logical possibility (and first-order logic) lets us formulate a version of Putnam's Potentialist set theory which differs from, and simplifies, Hellman's formulation. Then I'll provide a set of axioms for conditional logical possibility, arguing for the truth of these axioms and showing they justify mathematicians' use of the ZFC axioms.

6 Purified Potentialist Set Theory: An Informal Sketch

In this chapter I will informally present my preferred version of Potentialist set theory (using the notion of conditional logical possibility) and clarify some philosophical issues about it.

I will employ a version of Putnam's approach, but appeal to logical possibility specifically (much as Hellman does) rather than mathematical possibility. So, when I say that it would be possible to have an initial segment V, I will mean (something like) that it would be logically possible for the objects which satisfy "is a penciled point," "is connected by an arrow" to form an intended-width initial segment of sets when considered under these relations.[1] By using non-mathematical relations, we avoid having to presume there is an antecedently meaningful notion of set or other mathematical relation.

I'll now give the details of this proposal.

6.1 Two Sorted Initial Segment Structures

I will only require my iterative hierarchies to satisfy IHW, not ZFC_2 as Hellman does, for the reasons discussed in Section 3.2. Doing this makes it convenient to admit the levels in our hierarchies of sets as primitive objects in their own right rather than rely on the (non-obvious) fact that Von Neumann ordinals can serve that function inside the sets. So, my iterative hierarchies will have two kinds of (first-order) objects playing two different roles: those of sets and ordinal levels with sets being related to one another by elementhood, ordinal levels being related to one another by less than, and

---

[1] So, for example, although I may casually talk about the possible existence of initial segment structures V_i, I don't mean to assert that there are (or could be) special objects called structures, as e.g., Shapiro does. Or at least, I don't want to say that we need such objects to understand set theory. All talk about "the possibility of a structure existing" in the Potentialist paraphrase strategy above is merely shorthand for claims about the possibility of there being objects which instantiate specific non-mathematical first-order predicates and relations in a certain way.

every set being "available at" some ordinal level. To reiterate, on my current way of talking the ordinals are not themselves sets.[2]

Thus, I will employ five relations (any relations of the right arity will do) to characterize the notion of initial segment: two one place relations set() and ord() identifying the objects playing the role of the sets and ordinals, and three two place relations playing the roles of $\in$, $<$ (ordinal ordering) and @ ("is available at," where a set x is available at an ordinal s if it has been constructed at or before stage s).

So, for example, we might use the following first-order properties and relations: ... is a penciled point, ... is a penciled star, ... is connected to ... by a dotted/dashed/solid arrow.

I will define a formula $\mathcal{V}(\text{set}, \text{ord}, <, \in, @)$ which asserts the relations set, ord, $<$, $\in$, @ apply to a objects in such a way as to satisfy our conception IHW (described in Definition 2.1) of an initial segment of sets. See Definition A.2 in Appendix A for a fully formal definition. For brevity I will often simply call these initial segments.

6.2   Structure Preserving Not Object Preserving Extensibility

I will use the notion of conditional logical possibility to talk about how one hierarchy of sets like structure could extend another.

I will define a formula $V' \geq V$, which says that one initial segment extends another, where V abbreviates a list of relations set, ord, $<$, $\in$, @ and V' abbreviates set', ord', $<'$, $\in'$ @.[3]

Now we can say that it's logically possible for an initial segment V to be extended by an initial segment V' by simply holding fixed (the relations in) V. I adopt the following abbreviation for this frequently used expression:

$$\Diamond_V (V' \geq V) \overset{\text{def}}{\leftrightarrow} \Diamond_{\text{set,ord},\in,<,@} (V' \geq V)$$

This might be read as saying[4] that it's logically possible (holding fixed the structure of V) for another initial segment V' to extend V.

6.3   Assignment Functions and Hierarchies

To completely eliminate quantifying in, I won't just think about how initial segments can be extended, but rather how initial segments augmented with a "function" representing an assignment of variables to sets can be extended. Thus, rather than talking

---

[2] Of course, should one desire, one can prove that my ordinals can be uniquely identified with the sets forming the Von Neumann ordinals in my system.

[3] So $V' \geq V$ says that the objects satisfying set', ord', $<'$, $\in'$@ form an initial segment extending the initial segment formed by the objects satisfying set, ord, $<$, $\in$, @ .

[4] In even more detail, it might be read as saying: It's logically possible (given the structure of the pencil points and arrows etc.) that the pen points and arrows etc. form an initial segment extending an initial segment structure formed by the pencil points and arrows.

about what's possible given an initial segment V and the object bound to the variable x as Hellman would, I talk about what's possible given an initial segment V and an assignment function ρ, where $\rho(\ulcorner x \urcorner)$ is meant to capture the assignment of the variable "x."

In particular, I'll associate each initial segment V with a copy of the natural numbers $\mathbb{N}$ (shared between all initial segments[5]) and an assignment function ρ assigning numbers to sets in that initial segment. Call the resulting structure an interpreted initial segment (see Definition A.5 in Appendix A for a formal definition), and let $\vec{\mathcal{V}}(\vec{V})$ abbreviate the conjunction of the requirement that V is an initial segment, $\mathbb{N}$ is a copy of the natural numbers and ρ is a function from $\mathbb{N}$ to the sets in V. I adopt the convention of using variables such as $\vec{V}, \vec{V}'$ to abbreviate corresponding lists of relations $V, \mathbb{N}, \rho$ and $V', \mathbb{N}, \rho'$.

So, recall that to give a Hellman style Potentialist translation of a sentence like $(\forall x)(\exists y)\phi(x,y)$, where ϕ is quantifier free, we want to say something like this:

It's logically necessary that however one chooses a set x from an initial segment V, it's logically possible to extend this V with an initial segment V' containing a set y making $\phi(x,y)$ true.

We can capture the same content as the above sentence as follows:[6]

$$\Box[\vec{\mathcal{V}}(\vec{V}) \rightarrow \Diamond_{\vec{V}'}[\vec{V}' \geq_y \vec{V} \wedge \phi(\rho(\ulcorner x \urcorner), \rho'(\ulcorner y \urcorner))][\in / \in']]$$

Here $\ulcorner x \urcorner, \ulcorner y \urcorner$ are objects in $\mathbb{N}$[7] coding the variables "x" and "y." And $\vec{V}'$ extends another ensemble $\vec{V}$ except on "y" (written $\vec{V}' \geq_y \vec{V}$) says that the initial segment V' extends V, we hold $\mathbb{N}, S$ fixed, and the assignment function ρ' maps $\mathbb{N}$ to the sets in V' such that $\rho(n) = \rho'(n)$ for all numbers n *except* $\ulcorner y \urcorner$.

Ignoring the details for the moment, the key insight here is that the initial logical necessity operator lets ρ range over all possible relations, so the consequent must hold given any possible set (position) in V chosen by $\rho(\ulcorner x \urcorner)$.

So, the claim above says (in effect) any way that $\rho(\ulcorner x \urcorner)$ could choose an "x" in an initial segment V, it would be logically possible to freeze this choice, and have an extended interpreted initial segment $\vec{V}'$ with a ρ' assigning "y" (and keeping the existing assignment of "x") so that $\phi(x,y)$ holds between the objects assigned to "x" and "y" respectively.

6.4   The Final Product

My strategy will be to translate the set-theoretic sentence $(\exists x)(\forall y)[x = y \vee \neg y \in x]$ with a Potentialist claim about what is conditionally logically possible, given the structural facts about how some relations $set_1, \in_1, \ldots, \rho_1$ apply as follows.

---

[5] That is, held fixed by our conditional logical possibility operators when we talk about possible extensions.
[6] Note, here I use functional notation for ρ, i.e., I write $\rho(x) = y$ rather than $\rho(x,y)$.
[7] $\ulcorner x \urcorner$ is represented as $S(S(S(\ldots S(0)))$ for some number of successor operators and 0 is the unique element of $\mathbb{N}$ that isn't a successor and S is a relation that we write functionally.

With the definitions above in place we can give a translation of $(\exists x)(\forall y)$ $[x = y \lor \neg y \in x]$ as follows:

$$\Diamond(\vec{\mathcal{V}}(\vec{V}_1) \land \Box_{\vec{\rho}_1}[\vec{V}_2 \underset{y}{\geq} \vec{V}_1 \to p_2(\ulcorner x \urcorner) = p_2(\ulcorner y \urcorner) \lor \neg p_2(\ulcorner y \urcorner) \underset{2}{\in} p_2(\ulcorner x \urcorner)])$$

However, I'll make one final tweak to the strategy illustrated above, to allow us to treat quantifiers in a uniform fashion. In the above examples the first quantifier had to be treated in a special manner as (the relations abbreviated by) $\vec{V}_1$ were not required to extend any $\vec{V}_0$. To this end, our translations will introduce a $\vec{V}_0$ and insist that $\vec{V}_1$ extend $\vec{V}_0$. Thus, for example, my official translation of $(\exists x)(\forall y)[x = y \lor \neg y \in x]$ is actually:

$$\Box(\vec{\mathcal{V}}(\vec{V}_0) \to \Diamond_{\vec{V}_0}(\vec{V}_1 \underset{x}{\geq} \vec{V}_0 \land \Box_{\vec{\rho}_1}[\vec{V}_2 \underset{y}{\geq} \vec{V}_1 \to$$
$$p_2(\ulcorner x \urcorner) = p_2(\ulcorner y \urcorner) \lor \neg p_2(\ulcorner y \urcorner) \underset{2}{\in} p_2(\ulcorner x \urcorner)])$$

6.4.1   Recursive Definition of Potentialist Paraphrases

I will now describe recursive principles which let us translate every sentence in the first-order language of set theory into a claim about logically possible extensibility.

First, we define a partial paraphrase function t_n. Intuitively, $t_n(\phi)$ transforms a set-theoretic formula ϕ into a Potentialist claim about the possible extensibility of the structure V_n, where free variables are filled in by the assignment function p_n (coded by our assignment relation R_n). Informally, the idea here is that $t_3(\phi)$ says that it would be possible to extend the $\vec{V}_3$ structure in a way that makes the Potentialist translation of ϕ true relative to $\vec{V}_3$ (i.e., relative to the assignment of variables made by p_3 to sets in V_3). So, for instance, $t_3((\exists x)[x \in y \to z \notin x])$ will be the sentence

$$\Diamond_{\vec{V}_3}\left(\vec{V}_4 \underset{x}{\geq} \vec{V}_3 \land p_4(\ulcorner x \urcorner) \in p_4(\ulcorner y \urcorner) \to p_4(\ulcorner z \urcorner) \notin p_4(\ulcorner x \urcorner)\right)$$

Formally, the paraphrase function t_n is defined as follows.

Definition 6.1 (Potentialist Translation). For any number n and set-theoretic formula ϕ:

- $t_n(x_i \in x_j) = p_n(\ulcorner x_i \urcorner) \in_n p_n(\ulcorner x_j \urcorner)$
- $t_n(x_i = x_j) = p_n(\ulcorner x_i \urcorner) = p(\ulcorner x_j \urcorner)$
- $t_n(\neg \phi) = \neg t_n(\phi)$
- $t_n(\phi \lor \psi) = t_n(\phi) \lor t_n(\psi)$
- $t_n(\phi \land \psi) = t_n(\phi) \land t_n(\psi)$
- $t_n((\forall x)\phi(x)) = \Box_{V_n}[\vec{V}_{n+1} \underset{x}{\geq} \vec{V}_n \to t_{n+1}(\phi)]$

- $t_n((\exists x)\phi(x))$ is the claim that $\Diamond_{V_n}[\vec{V}_{n+1} \underset{x}{\geq} \vec{V}_n \land t_{n+1}(\phi)]$.

The translation of a set-theoretic sentence ϕ is $t(\phi) = [\mathcal{V}(V_0) {\to} t_0(\phi)]$.

In Definition 6.1 recall that $\Box_{\vec{V}_n}$ ($\Diamond_{\vec{V}_n}$) abbreviates a claim about what is logically necessary/possible holding fixed the facts about set$_n$, $\in_n$, ord$_n$, @$_n$, $\leq_n$, N, S, ρ_n.

In what follows, I will, consistent with our general policy for functions, write $\phi(\rho_n(\ulcorner x_i \urcorner))$ to abbreviate claims of the form $(\exists k)\rho_n(\mathbf{i}, k) \wedge \phi(k)$. Moreover, since I will always subscript all the relations ρ_n, N, S whenever I subscript ρ_n, I will assume that any $\Box$ or $\Diamond$ subscripting ρ_n actually subscripts ρ_n, N, S.

6.4.2   Equivalence of Approaches

It's useful to observe that my choice to add a base initial segment $\vec{V}_0$ into my translations is purely a matter of convenience as the two translation schemas turn out to be logically equivalent in my system.

So, for example, the straightforward paraphrase for $\exists x \phi(x)$ would be

$$\Diamond \left[\left(\mathcal{V}(\vec{V}_1) \wedge t_1 \left(\phi(x) \right) \right) \right]$$

and this is (in the formal system I propose) logically equivalent to my official paraphrase for $\exists x \phi(x)$:

$$\left(\mathcal{V}(V_0) {\to} \Diamond_{V_0} [\vec{V}_1 \geq \vec{V}_0 \wedge t_1 (\phi)] \right)$$

See Lemmas L.6 and L.7 in Section L of the online appendix for a proof of this equivalence.

6.5   Atomic Predicate Use Reducing Trick

One final question which naturally arises is whether this style of Potentialist paraphrase requires appeal to infinitely many atomic predicates. As stated so far, my strategy would require access to unboundedly many atomic relations if we want to be able to translate set-theoretic sentences with arbitrarily deep nested quantifiers. For instance, the Potentialist translation of $\forall x \exists y \forall z \phi(x, y, z)$ would seemingly require three distinct tuples of relations $\vec{V}, \vec{V}', \vec{V}''$.

However, a careful examination of our translations shows that we only preserve the relations from the prior logical possibility context. Thus, if desired, in the above Potentialist paraphrases we can replace V^n with $V^{n\bmod 2}$ (where V^1 is just V', V^2 is V'' etc.) without affecting the truth-value of the translation. This allows us to translate sentences with arbitrarily many quantifier alternations using a fixed finite number of atomic relations.

Here's what I mean. We translate a sentence with three quantifiers $\forall x \exists y \forall z \phi(x, y, z)$ as follows:

$$\Box\left(\vec{\mathcal{V}}(V)\to\Diamond_{\vec{V}}\left[\vec{V}'\geq_y\vec{V}\wedge\Box_{\vec{V}'}\left(\vec{V}''\underset{z}{\geq}\vec{V}'\to\phi(x,y,z)\right)\right]\right)$$

But note that logical possibility treats all relations of the same arity the same. And conditional logical possibility treats all relations (that aren't being held fixed) of the same arity the same. So, the assertion

$$\Box_{\vec{V}'}\left(\vec{V}''\geq_z\vec{V}'\to\phi(x,y,z)\right)$$

is true if and only if

$$\Box_{\vec{V}'}\left(\vec{V}\geq_z\vec{V}'\to\phi(x,y,z)\right)$$

That is, replacing V'' with V has no effect. So we can formalize the same claim as follows:

$$\Box\left(\vec{\mathcal{V}}(V)\to\Diamond_{\vec{V}}\left[\vec{V}'\geq_y\vec{V}\wedge\Box_{\vec{V}'}\left(\vec{V}\underset{z}{\geq}\vec{V}'\to\phi(x,y,z)\right)\right]\right)$$

For readability I'll write as if I have access to an unbounded number of distinct relations of each arity. But keep in mind that the argument above demonstrates we can limit ourselves to only 16 distinct atomic relations.

7 Content Restriction

Now let's turn to the question of how to reason about logical possibility. Before I can present my formal system for reasoning about logical possibility, I must first introduce a key concept: content restriction.

To crudely motivate the idea of content restriction, recall that the language of logical possibility $\mathcal{L}_\Diamond$ doesn't allow quantification into the logical possibility operator $\Diamond$. Doing this will be convenient for providing a formal system whose inference rules' correctness is easy to recognize (without soundness proofs that would be question begging our current context). For it (in effect) cleaves good reasoning about logical possibility into two parts:

- In one part we use standard first-order logic to reason about a given logically possible scenario/what an arbitrary logically possible scenario must be like.
- In another part we use special mdal-structural principles to establish which scenarios are logically possible, and transfer facts about one scenario to another.

To state intuitive principles of the latter kind, I'll specify a class of sentences whose syntactic form ensures that they *only talk about the structure of objects related by certain relations* – and hence must remain true in all logically possible scenarios which hold that structure fixed. I will say that such sentences are **content-restricted** to the relevant relations $R_1, \ldots, R_n$.

7.1 A Motivating Example

To make this point more concretely, consider purely number theoretic statements, i.e., statements whose syntactic form makes it clear that they only make a claim about the structure of the natural numbers, rather than a claim whose truth-value might reflect the behavior of some larger universe of objects. Intuitively, the truth-value of such statements is completely determined by the structure of the natural numbers. That is, their truth-values are completely determined by structural facts about how the relations $\mathbb{N}, S, +$ and $\times$ (S for "successor") apply (call this the $\langle \mathbb{N}, S, +, \times \rangle$ structure) and don't depend on what other objects may or may not exist.

A key strategy for finding suitably obvious modal structural principles will be to latch on to certain sentences whose syntax ensures that they *only talk about* a given structure, and hence can be assumed to preserve their truth-value in every logically possible scenario where this structure is preserved. I will say that such sentences are **content-restricted** to (the relations forming) the relevant structure.

So, for example, consider the claim that there are infinitely many twin primes. We know that if the current state of the world makes this statement true, then facts about the natural number structure alone suffice to do this. Because it only quantifies over the natural numbers, and only concerns itself with how the relations successor, plus and times apply to the natural numbers, this sentences' truth-value can't be changed by adding or subtracting objects from the universe outside the extension of $\mathbb{N}$ or by tinkering with the extension of properties and relations other than $S, +$ and $\times$ (like "is democratically governed" or "is a spaceship").

Accordingly, we expect that the truth-value of any purely number-theoretic statement cannot be changed by the application of any conditional logical possibility operator which holds fixed the natural number structure (i.e., the facts about how $\langle \mathbb{N}, S, +, \times \rangle$ apply). For, intuitively, modifying what other objects exist outside the structure of the natural numbers which this statement quantifies over (and/or changing the extension of other relations which it doesn't employ) can't affect its truth-value.

This idea (that truth-value of all purely number theoretic claims must be the same in all logically possible worlds which preserve the structures of the natural numbers) fits well with common intuitions about the significance of non-elementary proofs.

Thinking about how the natural numbers would relate to other larger mathematical structures, like the complex numbers, can be epistemically helpful in discovering the answer to some purely number theoretic statements. Proofs of this kind are called non-elementary proofs. But we don't think about the existence of the complex numbers as helping make these number-theoretic statements true or false. Rather we think that, if true, the relevant number-theoretic statements must have been true *all along*, just because of what the natural numbers are like. Considering how the natural numbers are (or could be) related to the complex numbers just helps us see this fact.

Accordingly, it doesn't matter to our acceptance of a non-elementary proof whether we think the complex numbers actually exist, or merely that it would be logically coherent for (an instance of) the actual natural number structure to exist inside (an instance of) the complex numbers. Showing the twin prime conjecture is true and merely showing that it would have to be true in the relevant logically coherent scenario both suffice to establish its truth in the actual world.

7.2   Generalizing This Idea

Generalizing the example from Section 7.1, we want to develop the idea that the syntactic form of certain sentences ensures they *only talk about a certain structure* (e.g., the natural number structure) – and hence must remain true in all logically possible scenarios which hold that structure fixed.

So consider what sentences we'd say obviously "only talk about" the natural number structure. The syntax of a sentence ϕ ensures that it only talks about the natural numbers if (not to say only if!) it has the following pair of features:

- All quantifiers in ϕ are restricted to the objects satisfying $\mathbb{N}$. Thus, it only contains universal quantifiers as part of expressions of the form $\forall x(\mathbb{N}(x) \rightarrow \psi)$ and existential quantifiers as part of expressions of the form $\exists x(\mathbb{N}(x) \wedge \psi)$.
- ϕ is a sentence in the language of number theory, so it only contains relations on this list: $\mathbb{N}, S, +, \times$.

If you accept the intuitions I've tried to pump here, you'll expect that ϕ cannot change truth-value in any conditionally logically possible scenarios which hold the facts about the natural numbers fixed. Accordingly, ϕ is actually true iff it is conditionally logically possible – holding fixed the natural number structure – that ϕ be true:

$$\phi \leftrightarrow \Diamond_{\mathbb{N},S} \phi$$

I will generalize this idea by considering other lists of relations $R_1, \ldots R_n$ (rather than just $\mathbb{N}, S$). I will define a syntactic property of sentences which intuitively ensures that a sentence is completely about (structural facts concerning) how some list of relations $R_1; \ldots; R_n$ apply, so that its truth-value (intuitively) must remain fixed in all conditionally logically possible scenarios which hold these relations fixed. I will call this property **explicit content restriction**. Accordingly, when a sentence ϕ is content-restricted to some list of relations $R_1; \ldots; R_n$, it is intuitively clear that $\phi \leftrightarrow \Diamond_{R_1 IR_n} \phi$

However, one little wrinkle arises in performing this generalization. In the case of the natural numbers, we thought of the structure of the natural numbers under the relations $S, +, \times$. And we said that purely number theoretic statements had their quantifiers restricted to objects in the extension of $\mathbb{N}$. But now we want to generalize this idea of only talking about the structure determined by how some arbitrary list of relations $R_1; \ldots; R_n$ apply. And this list of relations doesn't have one particular property distinguished as representing the domain.

So what exactly should it mean to talk about "the $R_1; \ldots; R_n$ structure"? Specifically, what domain of objects should we consider the behavior of, under the relations $R_1; \ldots; R_n$? I will handle this problem by (in effect) considering the domain of objects which *any one* of the relations $R_1; \ldots; R_n$ apply to, under the relations $R_1; \ldots; R_n$. So, for example, if our list of relations $R_1; \ldots; R_n$ is Cat(), Loves(), then an object is in $\text{Ext}(R_1; \ldots; R_n)$ iff it is *any one of* a cat or a lover or a beloved.

And we will consider the structure of objects determined by the relations $R_1; \ldots; R_n$ (analogous to the natural number structure, in the original case) to be the structure formed by considering the objects that at least one of $R_1; \ldots; R_n$ applies to, under the relations $R_1; \ldots; R_n$ (e.g., the objects that either "cat" or "loves" applies to, under the relations Cat() and Loves()). With just a little abuse of notation, I will call this domain of objects associated with the $R_1; \ldots; R_n$ structure the *extension* of the list of relations $R_1; \ldots; Rn$, and give the formal definition below.

Definition 7.1 (Definition of Ext). Let $\text{Ext}(R_1,\dots,R_n)(y)$ abbreviate the formula

$$\bigvee_{\substack{1\le i\le n\\1\le j\le l_i}} (\exists x_1),\dots,(\exists x_{j-1}),(\exists x_{j+1}),\dots,(\exists x_{l_i})R_i(x_1,\dots,x_{j-1},y,x_{j+1},\dots,x_{l_i})$$

where l_i is the arity of R_i and $\bigvee_{\substack{1\le i\le n\\1\le j\le l_i}} \phi_{i,j}$ indicates the disjunction $\phi_{i,j}$ over all

indicated values for i and j. Thus, $\text{Ext}(R_1,\dots,R_n)(y)$ is the formula asserting that some tuple $\vec{v}$ including y satisfies some $R_i(\vec{v})$.

So I will take the $R_1 I R_n$ structure to be the structure of objects in $\text{Ext}(R_1,\dots,R_n)$ under the relations $R_1 I R_n$. And I will define a syntactic property of **explicit content restriction**, such that the fact that a sentence ϕ is explicitly content-restricted to some relations $R_1 I.R_n$ intuitively insures that ϕ "only talks about" the $R_1 I.R_n$ structure so that its truth-value will be preserved in all conditionally logically possible scenarios which hold this structure fixed.

7.3   Formal Definition

So *explicitly content restricted* sentences are supposed have a syntactic structure which ensures that their truth-value is completely determined by the $\mathcal{L}$ structure (for $\mathcal{L}$ a certain list of relations), with the result that $\phi\leftrightarrow\Diamond_{\mathcal{L}}\phi$.

To motivate my definition, consider two examples. First, the truth-value of the sentence "Every lover is loved by someone," $\forall x(\exists y \text{Loves}(x,y)\rightarrow\exists z\text{Loves}(z,x))$, is completely determined by facts about the Loves() structure. This sentence only makes a claim about the structure of objects which are either lovers or beloveds, under the relation Loves.

We can see this by noting that it is logically equivalent to a sentence with quantifiers explicitly restricted to objects in Ext(Loves) and the considerations from Section 7.2 above apply. Thus:

$$(\forall x|\text{Ext}(\text{Loves})(x))[(\exists y|\text{Ext}(\text{Loves})(y))\text{Loves}(x,y)\rightarrow(\exists z|\text{Ext}(\text{Loves})(z))\text{Loves}(z,x)]$$

and we clearly cannot change the truth-value of the resulting sentence by:

- adding or subtracting objects which Loves() doesn't apply to from the universe;
- changing the extension of predicates and relations other than Loves().

So the truth-value of this sentence must be preserved in all scenarios which hold the Loves() structure fixed. Thus, we intuitively have

$$\forall x(\exists y\text{Loves}(x,y)\rightarrow\exists z\text{Loves}(z,x))\leftrightarrow\Diamond_{\text{Loves}}\forall x(\exists y\text{Loves}(x,y)\rightarrow\exists z\text{Loves}(z,x))$$
$$\leftrightarrow\Box_{\text{Loves}}\forall x(\exists y\text{Loves}(x,y)\rightarrow\exists z\text{Loves}(z,x))$$

In contrast, the truth-value of the sentence "Everything loves something," i.e., $(\forall x)(\exists y)(\text{Loves}(x,y))$, is not completely determined by the Loves() structure. For the existence of objects outside of this structure can make a difference to its truth-value. Specifically, it is logically possible that this sentence be true. But, given any world where this sentence is true, we can imagine a logically possible scenario which holds fixed the structural facts about how Loves() applies at this world, but makes this sentence false by containing an additional object which loves does not apply to. Thus, the truth-value of this sentence is not completely determined by the world's Loves structure. We might have $(\forall x)(\exists y)(\text{Loves}(x,y))$ but $\Diamond_{\text{Cat,Loves}} \neg (\forall x)(\exists y)$ $(\text{Loves}(x,y))$.

Roughly speaking, I will say that a sentence ϕ is explicitly content-restricted to a finite set (note the notions of sets aren't presumed in the object language merely used to in the meta-language and can be easily be eliminated[1]) of relations $\mathcal{L}$ iff only the relations from $\mathcal{L}$ are used in ϕ and every quantifier is restricted to range over elements that belong to some tuple in the extension of a relation in $\mathcal{L}$.[2] The definition below expresses this idea.

Note that I will frequently drop the braces and union symbols when talking about sets of relations. For instance, I will abbreviate the claim that ψ is content-restricted to $\mathcal{L} \cup R \cup \mathcal{L}'$ simply as ψ is content-restricted to $\mathcal{L}, R, \mathcal{L}'$ and write $\mathcal{L} = R_1, \ldots, R_n$ rather than $\mathcal{L} = R_1, \ldots, R_n$.

Definition 7.2 (Content Restriction). A sentence ϕ is **explicitly content-restricted** to a list $\mathcal{L}$ if it is a member of the smallest set S satisfying:

1. $\bot$ is in S;
2. if v_i, v_j are variables the formula $v_i = v_j$ is in S;
3. if $\vec{v}$ is a tuple of variables and $R_i \in \mathcal{L}$ then $R_i(\vec{v})$ is in S;
4. if $\psi \in S$ and $\rho \in S$ then $\neg\psi$, $\psi \vee \rho$, $\psi \wedge \rho$ and $\psi \rightarrow \rho$ are all in S;
5. if $\psi \in S$ and $\mathcal{L}$ is non-empty, then $\exists y(y \in \text{Ext}(\mathcal{L}) \wedge \psi)$ is in S;
6. if $\psi \in S$ and $\mathcal{L}$ is non-empty, then $\forall y(y \in \text{Ext}(\mathcal{L}) \rightarrow \psi)$ is in S (note this case is added only for illustrative purposes as technically $\forall$ is merely an abbreviation for $\neg\exists\neg$);
7. if $\phi = \Diamond_{\mathcal{L}'}\psi$, where ψ is a sentence and $\mathcal{L}' \subseteq \mathcal{L}$ then $\phi \in S$. Note that ψ need not be in S.

Clauses 2–6 in Definition 7.2 express the following basic idea: if a sentence employs only quantifiers which are restricted to $\text{Ext}(\mathcal{L})$ and relations in $\mathcal{L}$, then it makes a claim which is purely about the $\mathcal{L}$ structure, and its truth-value must be completely determined by this structure.

---

[1] Sets merely serve as a convenient way to talk about syntactic properties of formulas in the language of logical possibility. Since we only make use of finite sets there is no need for a set-theoretic meta-language (we could directly give a computable enumeration of allowed inferences).

[2] The intuitive conception outlined here only makes sense for sentences, but I will define a notion of content-restricted for formulas as well so we can keep track of whether the sentences built from them are content-restricted.

Clauses 1 and 7 liberalize this definition slightly, by allowing two other basic ingredients to figure in sentences which are content-restricted to $\mathcal{L}$.

Clause 1 allows sentences which are content-restricted to $\mathcal{L}$ to employ the logically false proposition $\perp$. This is motivated by noting that because there is no way to change $\perp$'s truth-value at all, there is no way to change it while holding fixed the facts about the $\mathcal{L}$ structure. So $\perp$ is intuitively content-restricted to any list of relations $\mathcal{L}$.

Clause 7 allows sentences which are content-restricted to $\mathcal{L}$ to employ claims about conditional logically possible given the facts about the $\mathcal{L}$ structure (or some part of it). Recall that $\Diamond_{\mathcal{L}'}\psi$ says that it's logically possible for ψ to be true, while holding fixed the structural facts about how the relations in $\mathcal{L}'$ apply. Accordingly, only structural facts about the relations in $\mathcal{L}'$ should be able to make a difference to its truth-value (remember that we don't allow quantifying into the $\Diamond$, so ψ cannot contain any free variables). So, when $\mathcal{L}'$ is a subset of $\mathcal{L}$, we can't change the truth-value of this sentence without changing how some relation in $\mathcal{L}$ applies.

To see how this definition applies, let's consider some examples. Let $\mathcal{L} = R, Q$ where R is a two-place relation and Q is a one-place relation. Then:

- $(\forall x)(\forall y)(x = y)$ is not content-restricted to $\mathcal{L}$;
- $(\exists x)(Q(x) \wedge K(x))$ is not content-restricted to $\mathcal{L}$;
- $(\forall x)[x \in \mathrm{Ext}(R) \rightarrow (\forall y)(y \in \mathrm{Ext}(R) \rightarrow [R(x,y) \rightarrow Q(y)]])^3$ (which is first-order logically equivalent to $(\forall x)(\forall y)[R(x,y) \rightarrow Q(y)]$) is content-restricted to $\mathcal{L}$;
- $\Diamond_R[(\forall x)(R(x,x) \wedge (\exists y)S(x,y))]$ is content-restricted to $\mathcal{L}$.

Also note the following consequences of Definition 7.2:

- If $\mathcal{L}$ is a sublist of $\mathcal{L}'$, then all formulae ϕ which are content-restricted to $\mathcal{L}$ are also content restricted to $\mathcal{L}'$.
- A sentence is content-restricted to the empty list $\mathcal{E}$ iff it is a truth-functional combination of unsubscripted $\Box$ sentences, $\Diamond$ sentences and $\perp$.

As you may have noticed, explicitly content-restricted sentences are generally long and unwieldy. This can be annoying when writing up proofs whose inference steps can only (strictly speaking) be applied to sentences which are content-restricted to some list $\mathcal{L}$. To avoid this annoyance, I make the following definition.

Definition 7.3 A formula ϕ is **implicitly content-restricted** to $\mathcal{L}$ if there is a sentence ψ explicitly content-restricted to $\mathcal{L}$ and $\phi \leftrightarrow \psi$ can be derived (using no assumptions) using only the first-order inference rules.

I will then frequently use the shorthand of applying rules which (strictly speaking) can only be applied to content-restricted sentences to implicitly content-restricted sentences – taking the work of using first-order logic to deduce the explicitly content-restricted form of a sentence before applying the relevant rule (and then transforming it back after applying the rule) for granted.

---

^{3} That is, $(\forall x)[(\exists k)(R(x,k) \vee R(k,x)) \rightarrow (\forall y)[(\exists k')(R(y,k') \vee R(k',y)) \rightarrow (R(x,y) \rightarrow Q(x))]]$.

7.4 Content Restriction and Potentialist Paraphrases

With this definition of content restriction in hand, we see that our definitions of Potentialist translations for set-theoretic sentences and formulas are often content-restricted in useful ways.

Lemma 7.1 *If $\phi, \theta_1, \ldots, \theta_n$ are formula in the language of set theory, then:*

1. *$t_n(\phi)$ is always content-restricted to $V_n, R_n, \mathbb{N}, \rho_n$;*
2. *if ϕ is a sentence, then $t(\phi)$ is content-restricted to the empty list;*
3. *for all i, j if $\vec{\mathcal{V}}(V_i), t_i(\theta_1), \ldots, t_i(\theta_n) \vdash_\Diamond t_i(\phi)$ then $\vec{\mathcal{V}}(V_j), t_j(\theta_1), \ldots, t_j(\theta_n) \vdash_\Diamond t_j(\phi)$.*

Proof. Claims 1 and 2 follow immediately from the definition of content restriction and our Potentialist paraphrases (repeated below). Claim 3 follows by a tedious, but simple, induction on proof length, where we transform the t_i version of a proof to the t_j version by replacing every instance of a relation in V_{i+k}, ρ_{i+k} with the corresponding relation V_{j+k}, ρ_{j+k} and noting that the result is still a proof. ∎

7.5 Isomorphism

There is one further definition that is important to present before we offer axioms for reasoning about conditional logical possibility and that is isomorphism between structures. Informally speaking, we can define the claim that two structures $R_1, \ldots, R_n$ and $R'_1, \ldots, R'_n$ are isomorphic modally, as saying that it's possible for a relation f to map the objects in (the extension of) one structure to those in (the extension of) the other structure. More formally, I give the following definition (here, as elsewhere, I use talk of functions as shorthand for the corresponding relations, see Section A.1 in Appendix A).

Definition 7.4 (Isomorphism). A relation f is an isomorphism of $\langle R_1, \ldots, R_m \rangle$ with $\langle R'_1, \ldots, R'_m \rangle$ (henceforth written $\langle R_1, \ldots, R_m \rangle \cong_f \langle R'_1, \ldots, R'_m \rangle$) if:

- f is a bijection of $\text{Ext}(R_1, \ldots, R_m)$ with $\text{Ext}(R_1, \ldots, R_m)$ (note that the domain of f may be larger than $\text{Ext}(R_1, \ldots, R_m)$ so long as it behaves appropriately on $\text{Ext}(R_1, \ldots, R_m)$);
- f respects the relations R_i, R'_i, i.e.,
$$(\forall x_1) \ldots (\forall x_m)[R_i(x_1, \ldots, x_n)) \leftrightarrow R'_i(f(x_1), \ldots, f(x_n))].$$

8 Inference Rules

In this chapter, I'll present a formal system for reasoning about logical possibility whose principles seem clearly true, and whose inference methods seem clearly truth preserving. I will define the consequence relation ⊢ by listing closure conditions in this chapter and the next and will take ⊢ to be closed under first-order consequence, e.g., as defined in Shapiro and Kissel (2018).

The following additional axioms govern reasoning with □ and ◊ in my formal system.

Note that I will generally, but not exclusively, adopt the convention that when used as meta-variables the capital Greek letters $\Phi, \Psi, \Theta, \Xi, Y$ are restricted to sentences while the lower-case Greek letters $\phi, \psi, \theta, \xi, \upsilon$ may be formulas or sentences. I will also follow the convention, standard in philosophical presentations of modal logic, of calling modal inference rules axioms, and presenting them in this form, I will call such principles axioms even when presented in the form of inference rules, (i.e., closure conditions for ⊢), rather than sentences that can be inferred at any point.

8.1 ◊ Introduction and Elimination

Axiom 8.1 (◊ Introduction). $\Theta \rightarrow \Diamond_{\mathcal{L}} \Theta$.

This rule captures the idea that what is actual must also be logically possible and, indeed, logically possible holding any structural facts about the actual world you like fixed.

This rule corresponds to rule T (sometimes written equivalently as $A \rightarrow A$) in the familiar modal system S5 (Garson 2016). Examples:

- "There are two cats" → "It is logically possible, given what cats there are, that there are two cats."
- "There are two cats" → "It is logically possible, given what dogs there are, that there are two cats."

Axiom 8.2 (◊ Elimination). If Θ is content-restricted to a list of relations $\mathcal{L}$, then $\Diamond_{L} \Theta \rightarrow \Theta$.

This rule expresses the idea that when θ is content-restricted to $\mathcal{L}$, the truth-value of θ is totally determined by the facts about $\mathcal{L}$. For instance:

- "It is logically possible, given what cats there are, that there are two cats" ⇒ "There are two cats."

- But **not**: "It is logically possible, given what dogs there are, that there are two cats" ⇒ "There are two cats."

Note that the second inference is not permitted by my rule because θ ("there are two cats") is not content-restricted to the list $\{dog(\cdot)\}$.

The next basic axiom expresses the intuition that if Θ is content-restricted to $\mathcal{L}$ then holding fixed relations not in $\mathcal{L}$ doesn't affect the logical possibility of Θ.

8.2 ◊ Ignoring

Axiom 8.3 (◊ Ignoring). Suppose Θ is content-restricted to $\mathcal{L} = R_1, \ldots, R_n$ and $S_1, \ldots, S_m$ are relations not among $R_1, \ldots, R_n$. Then $\lozenge_{\mathcal{L}'}\Theta \rightarrow \lozenge_{\mathcal{L}',S_1,\ldots,S_m}\Theta$.

Remember that when a formula is content-restricted to $\mathcal{L}$, its truth depends only on facts about $\mathcal{L}$. This axiom reflects this intuition by allowing one to ignore irrelevant facts.

We will see that the converse inference, from $\lozenge_{\mathcal{L},S_1,\ldots,S_m}\Theta$ to $\lozenge_{\mathcal{L}}\Theta$ is also provable from the basic axioms and inference rules in this chapter (in Lemma B.1). Examples:

- "It is possible, given what cats there are,[1] that every cat admires a distinct dog" → "It is possible given what cats and dolphins there are, that every cat admires a different dog."
- But **not**: "It is possible, given what cats there are, that there are exactly three objects" → "It is possible, given what cats and dolphins there are, that there are exactly three objects."
- The latter conditional cannot generally be assumed, because the claim that there are exactly three objects is not content-restricted to any list of relations.
- And **not**: "It is possible, given what cats there are, that every cat admires a distinct dog" → "It is possible, given what cats and dogs there are, that every cat admires a distinct dog."
- Here "every cat admires a distinct dog" is content-restricted to {cat, dog, admires}, but for this inference to be permitted, "every cat admires a distinct dog" would have to be content-restricted to a list that didn't include the relation dog(). Facts about how dog() applies can make a difference to the truth-value of "Every cat admires a distinct dog," hence requiring that we hold fixed facts about what dogs there are could make a difference to the satisfiability of this claim.

8.3 Simple Comprehension

Axiom 8.4 (Simple Comprehension). Suppose $R \notin \mathcal{L}$ and R doesn't appear in ϕ or Ψ. Then $\Psi \rightarrow \lozenge_{\mathcal{L}}(\Psi \wedge (\forall \bar{z})[R(\bar{z}) \leftrightarrow \phi(\bar{z})])$.

This axiom schema captures the idea that any way a formula applies to objects is a logically possible way for a relation to apply to those objects. The inclusion of Ψ

---

[1] Here, and in the future, I mean this as shorthand for given the structural facts about the objects falling under the relation in question.

under the $\Diamond_{\mathcal{L}}$ reflects the intuition that the relation could apply in the way ϕ does without altering how any other relations apply or changing which objects exist.[2]

Example: "If there is something which everyone loves, it is logically possible (given the facts about love) that there is something which everyone loves *and* happy() applies to exactly those individuals which love themselves."

8.4   Relabeling

Axiom 8.5 (Relabeling). If $R_1, \ldots, R_n$ are distinct relations that don't occur in $\mathcal{L}$, and $R'_1, \ldots, R'_n$ is a list of distinct relations such that each R'_i has the same arity as the corresponding R_i and none of the R'_i occur in $\mathcal{L}$ or Θ, then $\Diamond_{\mathcal{L}}\Theta \leftrightarrow \Diamond_{\mathcal{L}}\Theta$ $[R_1/R'_1, \ldots, R_n/R'_n]$.

Here $\Theta[R_1/R'_1, \ldots, R_n/R'_n]$ denotes the simultaneous substitution of R'_1 for R_1, R'_2 for R_2 and so on. Note that when we give a list of relations $R_1, \ldots, R_n$ we usually assume they are distinct relations (no R_i is actually the same relation as R_j for $j \neq i$), but we state it explicitly here for clarity.

This axiom schema expresses the idea that when evaluating claims about logical possibility only the arity of a relation matters. Thus, replacing some $R \notin \mathcal{L}$ with an unused relation $R' \notin \mathcal{L}$ of the same arity cannot change the truth-value of $\Diamond_{\mathcal{L}}\theta$.

Example: By substituting "sleeps" with "chews," we see "It is logically possible, given the facts about dogs and blankets, that every dog sleeps on a different blanket" if and only if "It is logically possible, given the facts about dogs and blankets, that every dog chews on a different blanket."

Note that as $\Box_{\mathcal{L}}$ is just $\neg\Diamond_{\mathcal{L}}\neg$ it is trivial to show that relabeling applies to $\Box$ claims as well as $\Diamond$ claims. We formalize this observation in the following lemma.

Lemma 8.1 (Box Relabeling). *If $R_1, \ldots, R_n$ are distinct relations that don't occur in $\mathcal{L}$, and $R'_1, \ldots, R'_n$ are distinct relations not equal to any R_i with the same arities as $R_1, \ldots, R_n$ that don't occur in $\mathcal{L}$ and aren't mentioned in θ, then $\Box_{\mathcal{L}}\theta \leftrightarrow \Box_{\mathcal{L}}\theta[R_1/R'_1, \ldots, R_n/R'_n]$.*

8.5   Importing

Axiom 8.6 (Importing). If Θ is content-restricted to $\mathcal{L}$ then $[\Theta \wedge \Diamond_{\mathcal{L}}\Phi] \rightarrow \Box_{\mathcal{L}}(\Phi \wedge \Theta)$.

This rule captures the idea that any true sentence Θ which only talks about how some relations $\mathcal{L}$ apply must remain true in any logically possible context that holds the $\mathcal{L}$ facts fixed.

---

[2] Technically, without changing the structure of the objects which are taken to exist at that "world."

8.6 Logical Closure

Axiom 8.7 (Logical Closure). If $\Theta \vdash \Phi$ then $\Diamond_{\mathcal{L}}\Theta \rightarrow \Diamond_{\mathcal{L}}\Phi$.

This rule captures the idea that logical inference is universally valid. Thus, if we can deduce Φ from Θ then a "scenario" in which Θ is true must also be one in which Φ is true. Note that when $\mathcal{L}$ is empty, this rule performs the work of both the necessitation and distribution rules used in systems like S5.

8.7 Cutback

Axiom 8.8 (Cutback). For any list of relations $\mathcal{L} = R_1, \ldots, R_m$, $[(\exists x)P(x) \wedge (\forall x | \text{Ext}(\mathcal{L})(x))P(x)] \rightarrow \Diamond_{\mathcal{L},P}(\forall x)P(x)$.

This axiom schema expresses the idea that if a predicate P applies to all the objects which relations in $\mathcal{L}$ apply to (and P applies to at least one thing), then it is logically possible to have a cut back universe which preserves how P and relations in $\mathcal{L}$ apply and contains no objects outside the extension of P.

8.8 Modal Comprehension

Our next axiom schema, Modal Comprehension, expresses a somewhat similar idea to Simple Comprehension (Axiom 8.4). Modal Comprehension expands on the idea behind Simple Comprehension by ensuring the logical possibility that some relation applies to exactly those objects picked out by some *modal* sentence. Informally, the idea is that Modal Comprehension lets us make inferences like the following:

Siblings: Holding fixed the facts about the relations Married(x,y) and Sibling(x,y), it is logically possible to have a relation $R(x)$ that applies to exactly those married individuals x with more siblings than their spouse.

Note that having more siblings than one's spouse has to be cashed out in terms of the logical possibility of a surjective but not injective map from their siblings to those of their spouse. On first glance, it would appear this would require passing x (the individual for whom we wish to compare their siblings to those of their spouse) into the logical possibility operator evaluating the possibility of such a pairing. However, our language of logical possibility does not allow this kind of quantifying in.

Instead, we do this by using a special, otherwise-unused, n-place relation Q to label and preserve a choice for an n-tuple of objects in Ext$(\mathcal{L})$. We say that it is possible (fixing the $\mathcal{L}$ facts) for R to apply in such a way that, necessarily (fixing the $\mathcal{L}, R$ facts), R only relates objects in Ext$(\mathcal{L})$ and however Q chooses a unique n-tuple of objects in Ext$(\mathcal{L})$ for consideration, R applies to this n-tuple iff a certain modal claim ϕ describing the behavior of $\mathcal{L}$ and Q is true. In this case, the relevant $\mathcal{L}$ is Married, Sibling and the modal sentence ϕ is $\Diamond_{\text{Married,Sibling},Q} (\exists x)[Q(x) \wedge (\exists y)\text{Married}(x,y)$ and $Z(\cdot, \cdot)$ is a surjective but not injective map from the siblings of x to those of y.

We can thus express informal claims like SIBLINGS with a sentence of the following form:

$$\Diamond_{\text{Married,Sibling}} \Box_{\text{Married,Sibling},Q,R}(\exists!x|Q(x)) \rightarrow$$

$$(\exists x)(Q(x) \wedge [R(x) \leftrightarrow x \in \text{Ext}(\mathcal{L}) \wedge \phi)]$$

Since it is possible for Q to apply to any single object x, the necessity operator above ensures that R applies to exactly those x which have more siblings than their spouse. Or, to put the point differently, if there was some x in the extension of one of Married, Sibling but not $R(x) \leftrightarrow \phi$ then, intuitively, it would be possible for Q to apply to such an x contradicting the assumption. With this motivation in place, I can now state the Modal Comprehension Schema.

Axiom 8.9 (Modal Comprehension). If:

R does not occur in $\mathcal{L}$, Ψ or ϕ
Q does not occur in $\mathcal{L}$ or Ψ
ϕ is content-restricted to $\mathcal{L}, Q$
then $\Psi \rightarrow \Diamond_{\mathcal{L}}(\Psi \wedge \Box_{\mathcal{L},R}[(\exists!\vec{x}|Q(\vec{x})) \rightarrow (\exists \vec{x}|Q(\vec{x}))[R(\vec{x}) \leftrightarrow \text{Ext}(\mathcal{L})(\vec{x}) \wedge \phi(\vec{x})]])$

where $(\exists!\vec{x}|Q(\vec{x}))$ means there is a unique tuple $\vec{x}$ satisfying Q. Note that we take $(\exists!\vec{x}Q(\vec{x}))[\varphi(\vec{x})]$ to mean that Q applies to a unique n-tuple of objects $\vec{x}$ and those objects satisfy (not necessarily uniquely) $\varphi(\vec{x})$.

8.9   Possible Infinity

Next, I propose the following axiom, asserting that it would be logically possible for there to be infinitely many objects.

Axiom 8.10 (Infinity). $\Diamond \Psi$ where Ψ is the conjunction of the following claims:

1. The successor of an object is unique: $(\forall x)(\forall y)(\forall y')[S(x,y) \wedge S(x,y') \rightarrow y = y']$.
2. Successor is one-to-one: $(\forall x)(\forall y)(\forall x')(S(x,y) \wedge S(x',y) \rightarrow x = x')$.
3. There is a unique object that has a successor and isn't the successor of anything: $(\exists!x \exists y)S(x,y) \wedge (\forall y) \neg S(y,x))$.
4. Everything that is a successor has a successor: $(\forall x)[(\exists y)S(y,x) \rightarrow (\exists z)S(x,z)]$.
5. Successor is anti-reflexive: $(\forall x)(\forall y)[S(x,y) \rightarrow \neg S(y,x)]$.

Note that by clause 1 the relation S is a function.

8.10   Possible Powerset

Axiom 8.11 (Possible Powerset). If F, C are distinct predicates, and $\in_C$ a two-place relation, then $\Diamond_F \mathcal{C}(C, \in_C, F)$.

Here $\mathcal{C}(C, \in_C, F)$ is the conjunction of the following claims:

- $(\forall x)\neg(C(x)\wedge F(x))$, i.e., the objects satisfying F and C are disjoint.
- $(\forall x)(\forall y)(x\in_C y\rightarrow F(x)\wedge C(y))$.
- $\Box_{C,\in_C,F}(\exists x)[C(x)\wedge(\forall y)((F(y)\wedge K(y))\leftrightarrow y\in_C x)]$, i.e., it's necessary that however some predicate K applies to some objects satisfying F, there exists a corresponding class C whose elements are exactly the objects which F applies to.
- $(\forall y)(\forall y')(C(y)\wedge C(y')\wedge\neg y=y'\rightarrow(\exists x)\neg(x\in_C y\leftrightarrow x\in_C y'))$, i.e., classes are extensional (no two members of C contain, in the sense of $\in_C$, the same elements).

Intuitively, this axiom schema says that it is always possible to add a layer of classes to the objects satisfying some predicate F. Note that $C(C, \in_C, F)$ is content-restricted to $C, \in_C, F$.

8.11 Choice

Axiom 8.12 (Choice). For all $n\geq 0, m > 0$ if I is an n-ary relation (where a 0-ary relation is assumed to be $\neg\bot$) and $R, \hat{R}$ are $n + m$-ary relations with $\hat{R}$ not appearing in Φ or $\mathcal{L}$ (nor equal to I, R), then

$$\Diamond_{\mathcal{L},I,R}\Phi\wedge\left[(\forall\vec{x})(\forall\vec{y})\left(\hat{R}(\vec{x},\vec{y})\rightarrow R(\vec{x},\vec{y})\right)\wedge(\forall\vec{x})[I(\vec{x})\rightarrow(\exists!\vec{y})\hat{R}(\vec{x},\vec{y})]\right]$$

This axiom schema captures the same intuition as the axiom of choice in set theory. It says that if every x satisfying I is related to some y by R, then (fixing I, R) another relation $\hat{R}$ can behave like a choice function selecting a unique such y for each x.

Note that in the case where $n = 0$ (when I becomes just a necessary truth) the axiom asserts the possibility of an $\hat{R}$ which applies to a unique m-tuple in the extension of R.[3] The utility of this special case is demonstrated in the following proposition.[4]

Proposition 8.1 (Simplified Choice). *Suppose $\hat{R}\notin\mathcal{L}, R\in\mathcal{L}$ (and $\hat{R}$ is not the same relation as R) then$(\exists\vec{x})(R(\vec{x}))\wedge\Phi\rightarrow\Diamond_{\mathcal{L}}(\Phi\wedge(\exists\vec{x})[\hat{R}(\vec{x})\wedge\hat{R}(\vec{x})\wedge(\forall\vec{y})(P(\vec{y})\rightarrow\vec{x}=\vec{y})])$*

Proof. This follows directly from the Choice Axiom (Axiom 8.12) by letting $n = 0$ as we regard the 0-ary relation I appearing in the Choice Axiom as $\neg\bot$. ∎

8.12 Possible Amalgamation

My final and least obvious (but I hope, on reflection, still very plausible seeming) principle says that (when certain special conditions are satisfied) we can have disjoint structures simultaneously witnessing the conditional logical possibility of extending some core structure in different ways.

---

[3] Thanks to Peter Gerdes for pointing out that I didn't need to state this as a separate axiom.
[4] Note that in Actualist set theory this fact is guaranteed by applying comprehension with parameters, but our version of comprehension doesn't allow this.

Speaking informally and using quantifying in, we might say this axiom captures the intuition that if the most general laws of logic permit a certain scenario $\phi(x)$ for each x in some logically possible I then they don't forbid the "disjoint union" of these scenarios.

Crudely speaking, this axiom takes us from the logical possibility (given some starting structure $\mathcal{L}$), of satisfying a certain formula $\phi(x)$ *for any single* x in a base collection of objects (those satisfying some I in $\mathcal{L}$), to the logical possibility of an expanded universe where *for every* object x satisfying I, there is a corresponding structure (indexed to this object x) within which a version of $\phi(x)$ is true.

For example, if we take I to be the predicate person($\cdot$) and the $\mathcal{L}$ to be the list person($\cdot$), childOf(x, y) Possible Amalgamation licenses claims like:

If, for any choice of a person, there could be (holding fixed the facts about people and parentage) as many ghosts as that person has children, then (holding fixed the facts about people and parentage) it could be that, for every person x, there are as many ghosts-haunting-x (disjoint from everyone else's ghosts) as x has children.

While this principle in some sense serves the same purpose as the set theorist's axiom of Replacement, it differs in a critical way from the Actualist's axiom of Replacement. Actualist Replacement acts as a closure condition on a single structure (the hierarchy of sets) and it's this aspect which makes its consistency non-obvious. However, the Amalgamation axiom merely asserts the *possibility* of a scenario in which this disjoint union is realized. This possibility is obviously logically possible in a way that assuming the existence of a single structure closed under Replacement is not.[5]

As before (with Modal Comprehension, Axiom 8.9), articulating this principle can seem to require quantifying in to the $\lozenge$ of logical possibility. However, we can use the same trick (involving an otherwise unused predicate Q) to get around it, as we did when formulating Modal Comprehension.

Axiom 8.13 (Amalgamation). If:

- $\mathcal{L}$ is a list of relations which contains the predicate I but not Q or $R_1 I R_n$;
- Φ is content-restricted $\mathcal{L}, Q, R_1, \ldots, R_n$ (where $P, R_1, \ldots, R_n$ and $\mathcal{L}$ share no relations);
- $\hat{R}_1, \ldots, \hat{R}_n$ are otherwise unused relations, such that if R_i is an n-place relation then $\hat{R}_i$ is an $n + 1$ place relation.

Let $\Psi(x)$ be the formula

$$\bigwedge_{1 \leq i \leq n} (\forall \vec{v}) \left(R_i(\vec{v}) \leftrightarrow \hat{R}_i(\vec{v}, x) \right)$$

---

[5] While earlier Potentialist approaches, e.g., Hellman (Geoffrey Hellman 1994a), did try to justify their uses of Replacement, I believe this axiom is more clearly guaranteed not to contravene the most general laws of logic.

asserting that $\hat{R}_i$ with x inserted into the last place behaves exactly the same as R_i. Let $\pi(x, y)$ be the formula

$$\bigvee_{\substack{1 \le i \le n \\ 1 \le j \le l_i}} (\exists z_1), \ldots, (\exists z_{j-1}), (\exists z_{j+1}), \ldots, (\exists z_{l_i}) \hat{R}_i(z_1, \ldots, z_{j-1}, x, z_{j+1}, \ldots, z_{l_i}, y)$$

which asserts that x appears in some tuple ending with y satisfying some $\hat{R}_i$

$$\Box_{\mathcal{L}} \; [(\exists! x | Q(x))(I(x)) \to \Diamond_{\mathcal{L},Q} \Phi] \to$$
$$\text{then } \Diamond_{\mathcal{L}} \; [(\forall x)(\forall y)(\forall y')[(\neg y = y' \wedge \pi(x,y) \wedge \pi(x,y') \to x \in \text{Ext}(\mathcal{L})] \wedge$$
$$\Box_{\mathcal{L}, \hat{R}_1 \ldots \hat{R}_n} [(\exists! x)(Q(x))(Q(x) \wedge I(x) \wedge \Psi(x)) \to \Phi]]$$

Remember that $(\exists! x | Q(x))(I(x))$ indicates that there is a unique x such that Q and that this x also satisfies I.

To see how this captures the intuition from the start of the section, note that (informally) the antecedent merely asserts that for each x satisfying I (with the value of x conveyed by Q) it is logically possible that $\Phi(x)$. The consequent, in turn, asserts that it's logically possible to have a single logically possible structure which can be broken up into disjoint "domains" M_x for each x satisfying I such that $M_x \models \Phi$. Note that M_x here is just a way of talking about the objects satisfying $\hat{R}_i(\cdot, \ldots, \cdot, x)$ and we capture talk of modeling by considering the logically possible scenario in which $R_i(y_1, \ldots, y_n)$ holds iff $\hat{R}_i(y_1, \ldots, y_n, x)$. In other words, this complex sentence merely expresses the relatively straightforward intuition that if we can index logically possible scenarios by I, then it's logically possible to have a disjoint union of scenarios witnessing all the logical possibility facts indexed by I.

9 Defense of ZFC

Recall from Section 6.4 that we have recursive principles which let us translate every sentence in the first-order language of set theory into a claim about logically possible extensibility.

In Appendix C I prove the following result, which establishes that my Potentialist translations preserve first-order derivability. Let $\vdash_{FOL}$ denote first-order consequence.

Theorem 9.1 *(Logical Closure of Translation). Suppose Φ, Ψ are sentences in the language of set theory and $\Phi \vdash_{FOL} \Psi$ then $t(\Phi) \vdash t(\Psi)$*

To justify mathematicians' use of the ZFC axioms it remains to show that our Potentialist translations of the ZFC axioms of set theory can be proved using the inference rules for logical possibility in Chapter 8.[1] Since this theorem shows that every first-order logical proof in the language of set theory can be reconstructed as a proof of the set-theoretic translations of the relevant claims in the language of logical possibility, this suffices to justify normal mathematical practice.

9.1   Bounded Quantification

First, we have some axioms (Foundation, Extensionality, Choice, Comprehension and Union), whose Potentialist translations follow fairly immediately from the following lemma about set-theoretic statements with bounded quantifiers. While statements with unbounded quantifiers must be translated in terms of modal quantification over possible initial segments, subformulas ϕ containing only bounded quantifiers (see Definition L.1 in Section L.3 of the online appendix for a formal definition) can be unpacked in a way that eliminates *further* appeals to modal quantification, i.e., if ϕ is bounded then $t_n(\phi)$ will be equivalent to a claim about $IsIVV_n$ that doesn't involve any conditional logical possibility operators.

Intuitively, the idea is that if x is a set in some initial segment V then, since no initial segment V' can add any sets below x, asking what sets elements are of x in V (i.e., relative to the membership relation $\in$ given by V) is the same as asking about sets that could be elements in x in some possible extension V' of V. As a result, if ϕ is a set-theoretic formula with only bounded quantifiers, then $t_n(\phi)$ is equivalent to formula constructed by

---

[1] I'm deeply indebted to Peter Gerdes for his help with the appendixes (in this text and online). Without his help in simplifying proofs and spotting gaps and errors, this project would not have been possible.

changing the relation $\in$ in ϕ to $\in_n$ (replacing free variables with their interpretation as we would do if translating a quantifier free formula. More formally we can say the following.

Lemma 9.1 (Bounded Quantifiers Lemma). *Suppose $\phi(x_0, ..., x_n)$ is a bounded formula in the language of set theory with only $x_1, ..., x_n$ free. If (V_n, ρ_n) is an interpreted initial segment (Definition A.5) then*

$$\phi^{V_n}(\rho_n(\ulcorner x_0 \urcorner), ..., \rho_n(\ulcorner x_n \urcorner)) \leftrightarrow t_n(\phi)$$

where ϕ^{V_n} is the result of replacing all occurrences of $\in$ with $\in_n$

See Section L of the online appendix for a proof of this lemma.

Note that in what follows I only sketch the proofs of these claims. Full detailed proofs of these claims are proved in Section M of the online appendix. In particular, I'll gloss over the details of the assignment functions and moving in and out of $\Diamond$ contexts to give the reader a sense of the core ideas used in each proof and refer the reader to the appendix for full proofs.

9.1.1 Foundation

Proposition 9.1 (Potentialist Foundation).

$$t((\forall x)[(\exists a)(a \in x) \rightarrow (\exists y)(y \in x \wedge \neg(\exists x)\,(z \in y \wedge z \in x))])$$

As x is the only unbounded variable appearing in Foundation (in light of Lemma 9.1, Bounded Quantifiers), we can think of the Potentialist translation of Foundation as simply asserting that it's necessary that in any initial segment V no set x in V contains an infinite $\in$ descending chain.[2] This claim is formalized in the following lemma.

Lemma 9.2 *If V is an initial segment and x is a non-empty set in V, then there is some $y \in x$ such that $(\forall z | z \in x)(z \notin y)$.*

Intuitively, this claim follows from the fact that we defined an initial segment so that the ordinals are well-ordered and in turn $\in$ must be a well-founded relation (x must contain some y that was built at the least ordinal level and thus y can't contain any other member z of x).

9.1.2 Extensionality

Proposition 9.2 (Extensionality). $t((\forall x)(\forall y)([[(\forall z \in x)(z \in y) \wedge (\forall z \in y)(z \in x)] \rightarrow x = y))$

Again, this claim essentially asserts the (obvious) fact that extensionality holds in any logically possible initial segment.

---

[2] That is, it's logically impossible, holding V fixed that there is a set x in V such that there is no $y \in x$ such that y has no elements in x.

9.1.3   Union

Proposition 9.3 (Potentialist Union). $\vdash t(\forall z \exists a(\forall y \in z)(\forall x \in y)(x \in a))$

Informally, this requires that, for every logically possible initial segment V and set z in V, there is a logically possible extension V' of V containing $\cup z$. To prove this, it is enough to note that $\cup z$ is contained in V. This follows by considering part 7 of the definition of Initial Segment (Definition A.2) and noting that all sets appearing in $\cup z$ occur at earlier ordinals than does z and hence $\cup z$ must occur in any initial segment that contains z.

9.2   Comprehension

Proposition 9.4 *(Comprehension). If $\phi(x, w_1, \ldots, w_n)$ is a formula in the language of ZFC with free variables $x, w_1, \ldots, w_n$. Then*

$$t(\forall z \forall w_1 \forall w_2 \ldots \forall w_n \exists y \forall x[x \in y \leftrightarrow (x \in z \wedge \theta)]).$$

In essence, this claim says the following.[3] Suppose if $\vec{V}$ is an interpreted initial segment and $z, w_1, \ldots, w_n$ are sets in V and Ψ is a first-order formula in the language of set theory. Then we can $(\lozenge_{\vec{v}})$ have an initial segment $\vec{V}' \geq_y \vec{V}$ containing a set y whose elements are exactly those $x \in z$ that $\theta(x, w_1, \ldots, w_n)$ is potentialistically true of.

By part 5 of Definition A.2 (Fatness), it is enough to show that it's logically possible for some relation R to apply to just those $x \in z$ that make the potentialistic translation of $\psi(x, w_1, \ldots, w_n)$ true and thereby show that there is such a set y in the initial segment V. This claim follows by application of Modal Comprehension (Axiom 8.9) to the Potentialist translation of $\psi(x, w_1, \ldots, w_n)$.

9.3   Moving to an Extension

Now let's turn to the axioms of Pairing and Powerset. Unlike the previous axioms verifying the truth of Pairing and Powerset requires showing the logical possibility of a non-trivial extensions of a given initial segment, i.e., pairs and powersets of sets in an initial segment V don't necessarily exist in V.

We will vindicate the Potentialist use of these axioms by showing that given a logically possible initial segment (and an assignment satisfying the preconditions for the Potentialist translation of these axioms), there is a logically possible extension (as guaranteed by the Proper Extension lemma in Section K of the online appendix) which witnesses the truth of the of these claims. Intuitively, this duplicates the

---

[3] That is, glossing over the fact that claims about the existence of sets in an initial segment are actually claims about the logical possibility of an interpreted initial segment where the interpretation picks out those sets.

Actualist set-theoretic intuition that, given parameters from V_α, the truth of the Pairing and Powerset Axioms is guaranteed by the existence of an extending $V_{\alpha+1}$.

9.3.1 Pairing

Proposition 9.5 (Potentialist Pairing). $\vdash t(\forall x \forall y \exists z (x \in z \wedge y \in z))$

Informally, this requires that for any logically possible initial segment V and sets x, y in V there is some logically possible extension V' containing a set z which has both x and y as members.

    Since it is logically possible to extend any initial segment to a strictly larger one (as proved in Lemma K.4 of the online appendix), we are guaranteed the logical possibility of a non-trivial extension V' for any V. We can then use Simple Comprehension (Axiom 8.4) to demonstrate the logical possibility[4] of a predicate applying just to x and y. By the assumption that V' properly extends V, we know that V' contains an ordinal larger than any in V. So, by part 5 of the definition of an initial segment (Definition A.2) there is a set z in V' containing x and y.

9.3.2 Powerset

Proposition 9.6 (Potentialist Powerset). $t(\forall x \exists y \forall z [(\forall w)(w \in z \rightarrow w \in x) \rightarrow z \in y])$

This claim can be proved in the same manner as Potentialist Pairing (Proposition 9.5). The only difference is that instead of applying Simple Comprehension (Axiom 8.4) to yield a predicate applying to some pair of sets, we use it to show the logical possibility of a predicate applying to all the sets that are members of x.

9.3.3 Choice

Proposition 9.7 (Potentialist Choice).

$$t(\forall x [\emptyset \notin x \rightarrow \exists f \phi(f, x)])$$

where

$$\phi(f, x) \leftrightarrow \forall a \in x \ (f(a) \in a)$$

Note that a choice function in the sense relevant to set theory is defined in terms of a set of ordered pairs. In this proof we rely on a lemma (Lemma M.6 in the online appendix), proved by iterating the reasoning we used to establish pairing, that given any function

---

[4] In the actual proof, as we can't quantify in, we must instead assume that the claim fails and use Simple Comprehension (Axiom 8.4) and Simplified Choice (Proposition 8.1) to infer the existence of Q which applies uniquely to a pair x, y witnessing the failure and then use Q with Simple Comprehension (Axiom 8.4) to build the predicate.

(in the sense of a relation) taking sets in V to sets in V, it's logically possible to extend V to a V' in which this function is realized by a set of ordered pairs.

With this lemma in mind, I embark on the proof. To prove this claim, it is enough to show that if V is an initial segment and x is a set in V with $\emptyset \notin x$ then it's logically possible to extend V to some initial segment V' containing a choice function f for x. Intuitively, this is sufficient because of the way translation works for existential statements.

We now establish the logical possibility (holding V fixed) of a relation $\hat{R}$ coding a choice function for x. Unsurprisingly, we build $\hat{R}$ by applying Choice (Axiom 8.12). We first build a relation R, defined using Simple Comprehension (Axiom 8.4), such that $R(a, b)$ holds just if $b \in a \in x$. We also build a predicate I (also defined using Simple Comprehension) that applies just to elements of x. We can then deduce the existence of a choice function (in the sense of a relation) $\hat{R}$ via Choice (Axiom 8.12). $\hat{R}$ has the property that it associates each $a \in x$ to a unique $b \in a$. The desired conclusion follows using Lemma M.6 in the online appendix to deduce the existence of a V' extending V containing a set of ordered pairs coding this choice function.

9.4   Amalgamation Axioms

The final two ZFC axiom schemas we need to vindicate (Infinity and Replacement) require using the Amalgamation axiom (Axiom 8.13).

Recall from Chapter 8, that the Amalgamation axiom asserts a kind of simultaneous possibility intuition. It says that we can, so to speak, take the disjoint union of any indexed collection of logically possible scenarios (when the possible scenarios are characterized so as to satisfy certain intuitive non-interference conditions).

The key to justifying both Replacement and Infinity will be to combine Amalgamation with a little trick about equivalence classes which I'll now sketch (see the online appendix for detail) to justify the following extensibility principle.

Suppose there is some logically possible initial segment V_x for each[5] x satisfying some predicate I. Then we can have a single initial segment V_Σ which extends an isomorphic copy of each V_x. Moreover, if each V_x extends some initial segment V_- then we can have V extend that same V_-.

First we deploy Amalgamation (Axiom 8.13) to infer the logical possibility of the "disjoint union" of initial segments V_x extending V_0. Then we construct the initial segment V using Possible Powerset (Axiom 8.11). We take the elements of V to be the equivalence classes of elements of the disjoint initial segments V_x under the equivalence relation induced by possible isomorphism of initial segments. Specifically, we'll consider z and y are equivalent if they represent the same set in different initial segments V_x and $V_{x'}$.[6] When the ordinals of each V_x are drawn from the same well-ordering (or they

---

[5] Of course, formally, we can't quantify in so we express the possibility of V_x by talking about what's logically necessary if some predicate Q applies to a unique x satisfying I.

[6] Note that this is true exactly when it would be logically possible for a relation R to isomorphically map an initial segment of V_x to an initial segment of $V_{x'}$ taking x to y. See Theorem K.1 in the online appendix for details on this part of the argument.

all extend some common initial segment) we can replace the equivalence classes with these common elements.

9.4.1   Replacement

The axiom schema of Replacement asserts that the image of a set under any definable function will also fall inside a set.

Proposition 9.8 (Potentialist Replacement). *Let θ be any formula in the language of ZFC whose free variables are θ, so that, in particular, B is not free in θ. Then*

$$t(\forall a \forall w_1 \forall w_2 \ldots \forall w_n [\forall x (x \in a \rightarrow \exists! y\ \theta) \rightarrow \exists b \forall x (x \in a \rightarrow \exists y (y \in b \wedge \theta))])$$

Speaking loosely (in terms of quantifying-in), we want to show that following. Given an initial segment V and a set a in V, if for every $x \in a$ it's logically possible that there is some initial segment V_x extending V and a set y_x in V_x making the Potentialist translation of $\theta(x, y_x)$ true, then it's logically possible to have a single set b in some $V_{\Sigma+1}$ extending V that contains all those witnesses. To construct $V_{\Sigma+1}$ we must first establish the logical possibility of an initial segment V_Σ containing sets y_x for each $x \in a$ and then show it is logically possible to extend it by one layer to collect those sets together in a single set b.

Now the truth of the Potentialist translation of a set-theoretic sentence is "absolute," unlike the notion of truth in a model. That is, as shown in Section L.2 of the online appendix), if $\vec{V}_i \leq_x \vec{V}_j$ and they agree on the assignments of x and y_x then the truth-values of $t_i(\theta(x, y_x))$ and $t_j(\theta(x, y_x))$ are the same.

So it is enough to ensure that some initial segment V_Σ extending V has the property that for each $x \in a$, V_Σ extends some $\hat{V}_x \cong V_x$ where V_x makes the Potentialist translation of $\theta(x, y_x)$ true for some set y_x in V_x. To establish the possibility of this V_Σ, we invoke the key reasoning from the start of Section 9.4, to go from the logical possibility of the initial segments V_x to the logical possibility of a V_Σ extending isomorphic images of them.

9.4.2   Infinity

Proposition 9.9 (Potentialist Infinity).

$$t((\exists x)[\emptyset \in x \wedge (\forall y \in x)(S(y) \in x)])$$

where $S(y)$ is $y \cup y$.

Verifying this claim requires showing that it's logically possible to have some initial segment $V_{\omega+1}$ containing a set x that is closed under successor. The notation here is suggestive, in that it will be enough to argue that it's logically possible to have an initial segment that corresponds to $V_{\omega+1}$ in the normal Actualist hierarchy. This initial segment contains a set x whose members are all sets in V_ω. It is easy to see that x

will be successor closed. The difficulties are two-fold. First, we must establish the existence of an infinite well-ordering (particularly one without a maximal element). Then we must argue that one could flesh out that well-ordering to an initial segment.

The first claim is proved in the Infinite Well-Ordering Theorem (Theorem J.1) located in Section J.1 of the online appendix. The idea behind the proof is as follows. We start with the successor function S from Infinity (Axiom 8.10) and use it to build our infinite well-order $\omega, <_\omega$. Intuitively, ω is just the smallest successor closed collection from dom S containing 0 and $<_\omega$ is the order induced by S. To overcome the difficulty of defining $<_\omega$ from S without quantifying-in, we invoke Possible Powerset (Axiom 8.11) so we can quantify over (objects coding) classes of elements in S. Since ω is the smallest successor closed collection we may, using Simple Comprehension (Axiom 8.4) define $<_\omega$ by $x <_\omega y$ just if there is a successor closed class containing y but not x. The reader should consult Section M.9 of the online appendix for the lengthy process of verifying each element of the definition of a well-order is verified and that the resulting well-order has no maximal element as well (as well as possessing several other desirable properties).

Having demonstrated the logical possibility of an infinite well-order, we argue for the logical possibility of an initial segment whose ordinals have height $\omega + 1$ (it is logically possible to add a single element to the well-order ω above all the prior elements). This part of the proof is formalized in the Fleshing Out Theorem (Theorem K.2 in Section K.6 of the online appendix) but boils down to arguing that, for any ordinal o in a given well-order $W, <$, it's logically possible to have an initial segment V_o whose ordinals include $W {\restriction}_{\leq o}, <$.

This is established by supposing the claim fails and considering the least ordinal o for which it fails. If that ordinal is a successor ordinal (the case where $o = 0$ is trivial) we use the Possible Powerset axiom (Axiom 8.11) to construct V_o from V_{o-1}. The difficult case occurs when o isn't a successor (i.e., is a limit). In this case we leverage the logical possibility of V_u for every $u < o$ via the reasoning at the start of 9.4 to construct V_o. This establishes the possibility of $V_{\omega+1}$, which contains he desired successor closed set.

Note that this informal sketch describes multiple appendixes full of formal work. For a full proof we invite the reader to consult the Infinite Well-Ordering Theorem (Theorem J.1) located in Section J of the online appendix.

Part III

In the last third of the book, I will discuss how the Potentialist set theory I have advocated in Parts I and II can be attractively developed into a larger philosophy of mathematics.

10 Platonism or Nominalism?

I have argued that the Burali-Forti concerns and anti-arbitrariness worries discussed in Chapter 2 motivate Potentialism, and thus Nominalism, about **set theory** (if not other areas of mathematics) in a way that should be of interest to philosophers of many different stripes. But what we should say about other types of mathematical objects? It seems unattractive to suggest that talk of, say, the real numbers should be understood in a completely radically different way from talk of sets. Mathematics appears to have a somewhat unified subject matter. However, I'll suggest this doesn't mean that accepting Potentialism about set theory forces one to take a similarly Nominalist position about other talk of mathematical objects.

In this chapter I'll discuss two ways the Potentialism about set theory advocated in Parts I and II of this book can be developed into a larger philosophy of mathematics that satisfies intuitions about the unity of mathematics. In Section 10.1 I'll describe how this proposal can be developed in a Nominalist fashion and in Section 10.2 I'll consider a more ontologically realist (neo-Carnapian in flavor) alternative.

10.1 Ontologically Anti-Realist Options

So, let's begin by noting two obvious Nominalist options for developing the Potentialist set theory in the first two parts of this book into a larger philosophy of mathematics. Importantly, the proposals in this chapter don't just take other branches of mathematics to resemble Potentialist set theory in avoiding commitment to mathematical objects. Rather, they honor intuitions about the unity of mathematics more fully, by suggesting that all areas of pure mathematics can be seen as the investigation of (what's allowed by) the laws of logical possibility. In particular, my version of Potentialism that all pure mathematical claims can be written as *pure* statements of logical possibility, i.e., ϕ or $\Diamond\phi$ claims, where ϕ is a statement in the language of conditional logical possibility (note that claims of this form don't hold anything fixed and thereby ignore all contingent facts about the actual world).

10.1.1 Set-theoretic Reduction

First, we could understand mathematical statements outside set theory by combining Potentialism about set theory with set-theoretic foundationalism. Bourbaki, etc. have shown that we can systematically identify mathematical objects of various kinds with sets. So, one approach to apparent talk of pure mathematical objects that aren't sets (e.g., apparent quantification over natural numbers) is simply to reduce these claims to set-theoretic statements in the usual way, and then apply the Potentialist translation strategy I've advocated in this book.

Call this approach reduction to Potentialist Set theory. This approach might seem to face a problem regarding handling second-order quantification over mathematical structures like the natural numbers. For, remember that my Potentialist paraphrase strategy only applies to first-order sentences. However, one can cash such second (and higher) quantification out in terms of quantification over sets of objects (themselves identified with sets) in the structure in question.

As my version of Potentialism doesn't commit one to the existence of any mathematical objects, neither does this approach. For it ultimately cashes out apparent existence claims about the natural numbers (and other mathematical structures) in modal terms which don't commit one to the existence of corresponding objects.

This approach is convenient for illuminating connections between different areas of mathematics. But it requires us to pick some way of identifying talk of various mathematical structures with set-theoretic talk, which can seem to introduce something arbitrary and inessential, as Benacerraf (1965) famously pointed out.

10.1.2   Modal If-Thenism

A different Nominalist approach to mathematics beyond set theory is modal if-thenism along the lines of Hellman (1994). The key idea is to interpret utterances which seem to quantify over mathematical objects as really making claims about what it's logically necessary that any objects satisfying certain axioms (articulating our conception of relevant pure mathematical structures) must be like. Specifically, a pure mathematical statement ϕ which appears to quantify over objects forming some pure mathematical structure S (other than the hierarchy of sets) will be logically formalized as asserting the conjunction of the following claims. It's logically possible for there to be some objects with and relations instantiating the relevant structure S. And it's logically necessary that if there are some objects with this structure then the version of Φ which talks about these objects is true.

Here is a more detailed example. Consider an arbitrary statement Φ in the language of arithmetic, e.g., the claim that there are infinitely many twin primes. The modal if-thenist might take the true logical form of this statement to be the following conjunction of claims:

- It's logically possible for there to be objects which have the intended structure of the natural numbers under successor, plus and times (when considered under some relations $\mathbb{N}, S, +, \times$):

$$\Diamond PA_\Diamond (\mathbb{N}, S, +, \times)$$

Here PA$_\lozenge$ denotes the categorical description of the natural numbers in the language of conditional logical possibility $\mathcal{L}$ provided in Section J.3 of the online appendix.[1]

- It's logically necessary that if there are objects with this intended structure, then they must also make Φ true (i.e., the version of Φ which is modified to talk about the relevant relations $\mathbb{N}, S, +, \times$ is true):

$$\square[\text{PA}_\lozenge(\mathbb{N}, S, +, \times) \to \Phi]$$

Note that here (in line with the Putnamian approach to Potentialism discussed in Chapter 2) the mathematical-looking relation names $\mathbb{N}, S, +, \times$ I am using are merely a mnemonic device, and we can deploy this paraphrase strategy using any non-mathematical relations with the right arity.

Also note that the level of truth-value Realism delivered by this approach depends on whether we have a categorical conception of all pure mathematical structures talked about in the sentence to be paraphrased. If we do have such a categorical conception (and this conception is stateable in second-order logic, and thus in the language of conditional logical possibility), we will get definite bivalent truth conditions. If not, we may not.

This difference tracks a widely accepted division between (what are sometimes called) algebraic and non-algebraic theories within mathematics. As *Stanford Encyclopedia* puts it, "Roughly, non-algebraic theories are theories which appear at first sight to be about a unique model: the intended model of the theory. We have seen examples of such theories: arithmetic, mathematical analysis ... Algebraic theories, in contrast, do not carry a prima facie claim to be about a unique [structure]. Examples are group theory, topology, graph theory ..." (Horsten 2019).

In what follows I will focus on non-algebraic structures, because they are generally considered to be the most philosophically problematic (Potter 2007) portion of mathematics. But see Hellman (1996) for some appealing thoughts about how to treat algebraic theories in terms of a logical possibility operator. It might be interesting to try to develop a foundation for category theory using the conditional logical possibility operator I've advocated here.

10.2 Ontologically Realist Options

In addition to the Nominalist approaches to expanding Potentialism about set theory to a general philosophy of mathematics (in a unified way) developed in Section 10.1, we can also take an ontologically Realist neo-Carnapian approach. I will develop such an

---

[1] More specifically, I am using PA$_\lozenge(\mathbb{N}, S, +, \times)$ to mean the conjunction of PA$_\lozenge$ as defined in Section J.3 of the online appendix and with the claim that the relations $+, \times$, playing the role of plus and times, satisfy the usual axioms, e.g., $\forall x \forall y x + S(y) = S(x + y)$.

approach in Chapter 15, but let me briefly foreshadow its main outlines and (claimed) advantages here.

On the view I will advocate, mathematicians' acceptance of axioms entailing existence assertions about complex numbers can change the meaning of their quantifiers, so as to make a sentence like, "there is a number which is the square root of -1" go from expressing a falsehood to expressing a truth. More generally, mathematicians can reliably form true beliefs by introducing any logically coherent axioms they like. So, we say that mathematical objects literally exist. However, mathematical knowledge is closely connected to knowledge of logical possibility, in that our access to facts about pure mathematical objects is unmysterious given knowledge of logical possibility. Talk of mathematical objects can be seen as having the "core job" of enabling study of logical possibility facts, in much the way we might say talk of cities and countries has the core job of helping us understand facts about people's political interactions.

These features help neo-Carnapian Realism about mathematical objects satisfy the unity of mathematics intuition evoked at the beginning of this chapter, and let this view duplicate the benefits of Nominalism with regard to access worries.

However, I will suggest that accepting the existence of mathematical objects outside set theory has some advantages. In particular, I'll argue that it helps avoid (certain forms of) classic indispensability arguments against mathematical Nominalism. It also helps honor Benacerraf's idea that we should treat notions that function similarly ("there is a number between 5 and 10" and "there is a city between NY and LA") in the same way. One might argue that the Burali-Forti paradox gives us special reason for overriding this norm in the case of set theory, but we should otherwise follow it.

One might also feel that rejecting mathematical objects outside set theory fits uncomfortably with Realism about non-fundamental objects in the special sciences (contracts, languages, peer groups, social clubs). So, if you don't favor an ultra-spare ontology generally (as I don't), there's a kind of unity argument for favoring neo-Carnapian Realism about mathematical objects over Nominalism.

One might fear that accepting mathematical objects, but not sets, prevents set theory from doing the job mathematicians initially wanted it for: enabling comparative study and theorem transfer between different areas of mathematics. However, this is not the case, because we can use Potentialist set theory with ur-elements (or just conditional logical possibility directly) to do that job.

10.3   Agenda

In this final part, I'll discuss each of the just discussed ways of extending the Potentialist set theory developed in Parts I and II to a larger philosophy of mathematics (modal Nominalism and neo-Carnapian Realism) in some detail.

First, I'll consider Nominalism and the most well-known objection to general mathematical Nominalism: the Quinean indispensability argument (that we can't avoid quantifying over mathematical objects in formalizing our best scientific theories). I'll argue that Nominalists who are motivated by considerations about set theory presented in this book (rather than, e.g., empiricism/physicalism) can use the logical possibility operator plus some cheap tricks to answer classic Quinean indispensability arguments, but the reference and grounding based versions of the Quinean indispensability argument pose more of a problem for them.

Then, I'll turn to neo-Carnapian Realism about non-set-theoretic mathematical objects, and argue that adopting this option lets us evade or reduce the lingering indispensability worries from Section 10.2, while maintaining many of the benefits of Nominalism. I'll then use the logical possibility operator to develop the general neo-Carnapian picture in certain ways: proposing a dynamics for neo-Carnapian knowledge by stipulative (re)definition and a framework for evaluating meta-semantic answers to access worries.

I'll conclude by noting how both Nominalist and neo-Carnapian Realist philosophies of mathematics developed in this part of the book support traditional Structuralist and (to a certain extent) Logicist intuitions about the nature of mathematics.

11 Indispensability

11.1 Introduction

With these options for slotting Potentialist set theory into a larger Platonist and Nominalist philosophy of mathematics in mind, let's turn to the famous Quine–Putnam (Quine 1961) indispensability argument against mathematical Nominalism (some variants of which will also have force against neo-Carnapian realism about mathematical objects). Although I won't ultimately advocate Nominalism, clarifying whether the Nominalist can answer this and related indispensability challenges will help us choose between the Nominalism and neo-Carnapian realist options mentioned in Chapter 10. Doing so will also reveal some interestingly different roles mathematical objects can play in the sciences and an indispensability worry that applies equally to the neo-Carnapian realist and the Nominalist.

The classic Quine indispensability argument belongs to a broader family of related challenges. I'll try to clarify what's required for the Nominalist to adequately answer a (classic Quinean) indispensability challenge. I'll also highlight Grounding and Reference indispensability challenges and argue that these are usefully distinguished, both from each other and the classic Quinean indispensability challenge.

11.2 The General Form of Indispensability Arguments

Abstractly, as Colyvan (2019) suggests, it can be helpful to think in terms of a family of indispensability arguments, with the following shared form:

- We ought to have ontological commitment to all and only the entities that are indispensable to our best scientific theories.
- Mathematical entities are indispensable to our best scientific theories.
- We ought to have ontological commitment to mathematical entities.

Different specific indispensability arguments correspond to different versions of the claim that mathematical objects are indispensable to our best scientific theories. Most famously, we get the classic Quinean Indispensability argument by cashing out "indispensability" in terms of quantification and literal statement. This result in the following challenge:

Quinean (Literal Statement) Indispensability Challenge: How can we literally state our best scientific theories without quantifying over (and thus committing ourselves to) the existence of mathematical objects?

Recall that Quine's criterion says (in slogan form) that we are committed to believing in all the objects which a theory we believe in quantifies over. To cash out the slogan, consider any logically regimented theory T.[1] If this theory T logically entails $\exists x F x$ then anyone who accepts T is committed to believing in the existence of some objects satisfying F. Thus, accepting a theory with first-order existential quantification over Fs yields ontological commitment to Fs.[2]

If all this is true, then philosophers who deny the existence of some kind of objects F face a burden to provide (or at least make it plausible that one could in principle provide) a logical regimentation of their best total theory of the world which doesn't "quantify over Fs," i.e., doesn't imply that $\exists x F(x)$. Thus, it has been argued that the Nominalist about mathematical objects owes a logical regimentation (which I will sometimes call a paraphrase) of their best total theory which doesn't quantify over mathematical objects. I will say more about what it means to adequately capture the content of this theory in a way that doesn't use mathematical objects below.

Philosophers pressing this classic indispensability argument maintain that one can't adequately logically regiment certain key scientific theories Nominalists tend to accept without quantifying over mathematical objects. Standard textbook presentations of these theories seem to involve quantification over numbers, and it is not clear how to eliminate this. For example, consider the famous "inverse square" law relating mass, distance and gravitational force:

$$F = m_1 m_2 / r^2$$

It can seem far easier to logically regiment a theory including this law in a Platonist way (e.g., in terms of functions from physical objects to real numbers or a mass ratio relation between physical objects and real numbers) than in a Nominalist way.

I take this challenge to be widely accepted as, if not inescapable, something which has enough intuitive force to require an answer. One can think of it as arising from a default presumption that you should be able to say what you mean literally (understood here to require formalization in a logically regimented language[3]) together with

---

[1] Assume T is formulated in the language of first-order logic or some extension of it which adds other notions like a model necessity or possibility operator.

[2] I leave aside the vexed topic whether accepting a theory which lets one derive some second-order existence claim $\exists XX(c)$ commits one to the existence of second-order objects, as it won't matter for present purposes.

[3] One might think of this demand arising from a demand to provide a (logically regimented) Carnapian explication which solves puzzles and will stand up to arbitrary pedantic questioning literally and in a regimented language, plus the idea that if we apply Quine's criterion we get only commitment to objects that exist.

the appearance that it's impossible to thus literally state certain parts of widely accepted scientific theory without quantifying over mathematical objects.

While there is an obvious intuitive pull to the literal statement demand, some philosophers of mathematics have rejected it, and hence the Quinean indispensability challenge (Melia 1995; Azzouni 2003). They've noted that scientists often convey serious theories of what reality is like by speaking about what would be true under assumptions which they don't believe in (e.g., infinitely deep water, frictionless planes, ideal gasses). And they've used the role of such clear fictions in the sciences to reject the literal statement demand – and hence avoid admitting the existence of mathematical objects while accepting Quine's criterion. So, they allow that the literal truth of the theories they state when doing science would require the existence of mathematical objects, but they deny that these theories are (literally) true.

The persuasiveness of this response is a matter of significant controversy.[4] However, even if this response succeeds, it doesn't protect against the following explanatory indispensability worry (raised by Baker 2005) and advocated in works like Colyvan (2019), which Platonists have pressed in response:

Scientific Explanatory Challenge: Demonstrate that we can *explain* scientific facts without reference to mathematical objects whose existence we don't believe in.

We can fit this explanatory indispensability argument into the general form given at the start of Section 11.2, if we understand mathematical objects to be indispensable to a theory (literally stated or not) if their existence is required for ideal/adequate explanation of the phenomena which the theory is supposed to explain along the lines the theory suggests. Thus, merely rejecting demands for literal statement doesn't get Nominalists off the hook as regards indispensability worries as a whole (and below I will suggest two additional indispensability challenges that arise).

11.3   Answering Indispensability Arguments

Now what does it take for a philosopher to adequately answer an indispensability challenge?

11.3.1   Motivation via Specific Problem Cases

It might seem that, to answer the Quinean indispensability worry, the Nominalist must show how to plausibly logically regiment their best scientific theories without quantifying over mathematical objects. However, taken literally, this requirement unfairly stacks the deck against the Nominalist. For, plausibly, independent philosophical puzzles in metaphysics and the philosophy of science, physics, biology, etc. currently prevent *everyone* (Nominalist and Platonist alike) from attractively logically regimenting certain

---

[4] For example, see Colyvan (2010) for one influential argument against such an "easy road" response to the classic Quinean indispensability argument.

parts of our best total scientific theory. Indeed, in some cases it seems clear that accepting the existence of mathematical objects *couldn't possibly help* clear the roadblocks to attractively logically regimenting a certain kind of physical theory. Mathematical Nominalists' failure to solve these puzzles shouldn't count against them.

Instead, I take it, indispensability worries only create a serious challenge for a mathematical Nominalist because, and to the extent that, philosophers pressing access worries have highlighted specific portions of our best scientific theory, such that one of the two following conditions holds:

- We can currently see how to attractively platonistically paraphrase *this portion* of our total scientific theory, but not how to nominalistically paraphrase it.
- We have some positive argument (such as Putnam's counting argument to be discussed in Chapter 14) that no Nominalist theory can adequately logically regiment this portion of our total theory.

If the Nominalist can address all known specific indispensability arguments (by either providing a paraphrase strategy that lets one adequately capture the content of the disputed portions of our best scientific theory or blocking/refuting the relevant specific arguments), then they will also count as sufficiently diffusing Quinean indispensability worries[5] (at least for the time being).

11.3.2 Adequate Paraphrase, Craig's Theorem and Expressive Power

Second, we can ask, what does it take for a Nominalist (or Platonist) logical regimentation to literally state a natural language scientific theory? What does a logical regimentation of a part of our best scientific theory (in response to a classic Quinean challenge) need to do?

First, it's traditional and appealing to think that nominalistic logical regimentations provided in response to the Quinean challenge should be something which a human being could believe and assert. This gives rise to an expectation that a single natural language theory should be paraphrased by a single/finite collection of sentences[6] in a finite human-learnable language.

Second, we might want our nominalistic physical theory to capture the *inferential role* of our best scientific theory in combining with other claims we might learn are true to derive concrete consequences, e.g., whether a cannon ball will land before a feather, and the like. Thus we might want a general paraphrase *strategy* which can formalize both our scientific theory and a range of other scientific and observational statements *S* which can be used to derive statements from that theory – not just a logical regimentation of our best scientific theory alone. The human graspability idea – that the answer to a Quinean challenge should be something we (finite creatures speaking a human learnable language) could actually say in response to a demand – also

---

[5] In doing this they will be dispelling the apparent reasons for thinking that accepting Nominalism would prevent us from attractively logically regimenting our best theories.

[6] However, some Nominalists have also allowed paraphrases to be infinite sets of sentences if there is some schema or algorithm for unpacking them (Field 1980).

motivates a uniformity expectation. The Nominalist should be able to systematically unpack and explain what they really mean by their natural language scientific theory. Thus, one might expect that there should be a computable procedure that generates a nominalistic logical regimentation from the original English statement, when the original is sufficiently clear[7].

A third and, I think, crucial and under-emphasized desideratum, is that our nominalistic formalizations of scientific theories should be *able to express the kinds of constraints on non-mathematical reality which we intuitively expect the scientific theory being paraphrased to express.*

To illustrate this point, consider Craig's theorem. Craig's theorem might seem to immediately answer Quinean indispensability worries by showing that we can always transform a Platonist theory into a nominalistic theory that has all the same nominalistic consequences. For, as *Stanford Encyclopedia* puts it (Colyvan 2019):

[Craig's theorem] states that relative to a partition of the vocabulary of an axiomatizable theory T into two classes, t and o (theoretical and observational, say) there exists an axiomatizable theory T^* in the language whose only non-logical vocabulary is o, [which implies] all and only the consequences of T that are expressible in o alone. If the vocabulary of the theory can be partitioned in the way that Craig's theorem requires, then the theory can be re-axiomatized so that apparent reference to any given theoretical entity is eliminated.

However, it is generally agreed that Craig's theorem does not suffice to block Quinean indispensability worries. Why? People sometimes say that such theories are inelegant and unexplanatory. For example, Field (1980) raises the point about Craig's theorem:

[S]ince I don't know any formal conditions that would rule out such formal trickery, let me simply say that by "theory" I mean a reasonably attractive theory. "Theories" [like the ones we'd get by applying Craig's theorem] are obviously uninteresting, since they do nothing whatever towards explaining the phenomenon in question in terms of a small number of basic principles.

I agree that the type of nominalistic paraphrases that are ensured to exist by Craig's theorem can fail to answer important indispensability Quinean indispensability type worries, through failing to be explanatory.

However, there's another important further way Cragian paraphrases can fall short. Even though these paraphrases will logically imply all the same sentences involving purely nominalistic relations (i.e., relations which the Nominalist and Platonist agree necessarily don't apply to/relate any mathematical objects), they can fail to answer indispensability worries because the nominalistic vocabulary used by a Platonist regimentation of a theory is too impoverished to express the intuitive content of the scientific theory being logically regimented.

---

[7] So, for example, one (intuitively) can't logically regiment one's pure mathematical language (in response to indispensability worries) by just saying that all (of what would normally be considered) true Platonistic mathematical statements are to be considered as abbreviating tautologies and all false ones as abbreviating contradiction. One must instead provide some concrete algorithm for formalizing Platonistic sentences into nominalistic ones in a way that has this property.

Consider a straightforward Platonist logical regimentation of a physics textbook theory which makes lots of predictions about the (say) position, mass, charge, etc., of physical objects by first-order logically implying many claims about how certain relations between objects and numbers apply (e.g., a mass in grams relation $M(x, y)$ relating objects to real numbers or a mass ration relation relating pairs of objects to real numbers). As these predictions are not nominalistic consequences in the sense above, our Craigian re-axiomatization of this Platonist theory need not preserve them. It might not make any predictions about objects' mass, charge, etc. at all![8]

Accordingly, nominalistic regimentations of a theory (including those with the good feature guaranteed by Craig's theorem) can fail because they are *expressively inadequate*, even if they aren't *explanatorily bad*. A Nominalist paraphrase produced by applying Craig's theorem to the Platonist formalization of a scientific theory can fail to constrain physical reality[9] in the way that the Nominalist takes the natural language scientific theory to (even if this Craigian translation is unified and explanatory so far as it goes[10]). Therefore, Craig's Interpolation theorem does not, on its own, provide a satisfactory answer to Quine's literal statement challenge.

I will say that an **adequate** paraphrase strategy for some chunk of scientific practice is an algorithmic function which maps a collection of sentences S (sufficient to rationally reconstruct the practice in question) to a collection of formal sentences in some human learnable language L, in a way that preserves the truth-conditions for these sentences.

In the rest of this book, I will consider a Nominalist paraphrase strategy that produces logical regimentations which Platonists are forced to acknowledge as expressively adequate in the following sense. We can prove from metaphysical principles the Platonist accepts, that applying this paraphrase strategy to a scientific sentence Φ produces a nominalistic translation $T(\Phi)$, which is true at exactly the same metaphysically

---

[8] One might try to address this problem by saying that really the physical theory should be considered alongside a theory that connects physical statements to some kind of more concretely observable claims (not to say a sense data language). And one might hope that the latter predictions will ultimately be cashed out in terms that don't involve any quantification over mathematical objects, e.g., "the left side of the balance scales will be lower down" or "you will see a black dot" would be in nominalistic terms. If you knew this, then you'd know that applying Craig's theorem would give you a theory that at least got these observational consequences right.

However, there are two problems with this response. First, it's not clear that the Platonist ever has to cash out their observational predictions in nominalistic vocabulary (appeals to numbers seems natural and helpful for capturing the detail of what we can observe or predict we'll observe, e.g., for capturing different shades of colors or size we can observe or expect to observe). Second, and more importantly, however, the intended content of a physical theory that the Platonist and Nominalist alike will want to capture with their logical regimentations will generally go far beyond such observational predications. The Platonist and Nominalist alike want to state theories that tell us about invisible magnetic fields and remote stars and events in the ancient past etc.

[9] That is, this Craigian translation can rule out many of the metaphysically or epistemically possible scenarios which the Nominalist takes this scientific theory to rule out.

[10] Imagine a case where the physics textbook which is being nominalized via Craig's theorem includes some historical generalizations and claims about physicists, which allow straightforward nominalization. In this case the nominalistic consequences of the Platonistically regimented theory may be all and only the historical claims included and implied by it. So the nominalistic formulation of the explanation for *these claims* might be just as good as the Platonistic one.

possible worlds as Φ. Accordingly, if the Platonist regimentation gets the truth conditions for the relevant physical theory right then so can the nominalistic regimentation.

In what follows, I'll call a logical regimentation strategy that captures the intended (possible worlds) truth conditions a scientific theory would have *if* Platonism was true a **Platonistically acceptable paraphrase strategy**, and a Platonistically acceptable paraphrase strategy that doesn't quantify over mathematical objects a **Nominalistically acceptable paraphrase strategy**. Note that a Platonistically acceptable paraphrase generally won't look adequate from the Nominalist point of view. For it is likely to pair scientific sentences with sentences that imply the existence of mathematical objects, and hence are false at all possible worlds![11]

11.4   Other Indispensability Worries

11.4.1   Reference and Grounding Worries

In addition to the Quinean and Explanatory Indispensability problem, there are two other indispensability challenges facing the Nominalist. I will call these the Finitary Reference and Grounding challenges.

Let's begin with the Finitary Reference explaining challenge.[12] This challenge for the Nominalist concerns accounting for our claimed ability to use sentences in a finitely learnable language to draw the kinds of distinctions we take ourselves to draw.

Finitary Reference Challenge: Explain how your sentences (including those you take to be false) are able to have the (possible worlds) truth conditions which you take them to have. How are you able to finitely learn a language which can draw the distinctions which take your language to draw?[13]

---

[11] In some cases (such as the logical regimentation of physical magnitudes statements to be discussed in Chapter 14) it will be controversial whether a given Platonist/Nominalist logical regimentation is adequate in the sense described in this chapter. For, it can be philosophically controversial whether various non-mathematical objects invoked in the paraphrase exist and/or what truth-value the scientific sentences to be paraphrased take on at this possible world. In this case, I will say that the Platonist/ Nominalist can **attractively** paraphrase a theory to the extent that the additional philosophical commitments (outside of the existence/nonexistence of mathematical objects) are attractive.

[12] I have (speaking somewhat loosely) called this a referential indispensability worry, because it concerns our ability to "refer" to certain sets of supposedly possible worlds by uttering sentences which are true at exactly these worlds.

[13] We might intensify this reference explaining challenge by adding the following requirement: Explain how you *would still have* been able to finitely learn a language that can state a certain range of thoughts if the actual world had been different in certain ways!

Let me clarify this amplified challenge by making things more concrete. A nominalistic paraphrase strategy might suffice to answer the plain reference explaining challenge (by showing how creatures like us could form sentences that pick out suitable sets of possible worlds, e.g., those where one stick is exactly π times longer than another) but fail to answer this amplified reference explaining challenge as follows. This paraphrase strategy might explain our actual reference abilities only by exploiting certain contingent facts about the world. But if the Nominalist thinks that our having these referential abilities (e.g., our ability to mean "One stick is exactly π times longer than another") *isn't* hostage to these contingent facts about the actual world, then this more ambitious two-dimensionalist reference explaining challenge won't be solved.

Additionally, Nominalists face a grounding worry. One might worry that mathematical objects play an indispensable role in grounding the truth of applied mathematical statements. For example, one might ask the Nominalist, "What metaphysically grounds facts of the form 'This object is r times more massive than that one' if not a three-place relation (assigning pairs of objects to their mass ratio) between physical objects and numbers?" Accordingly, one might argue that Nominalists can't meet the following Grounding challenge:

Grounding Challenge: Explain how the truth of propositions you think are true in actual world can be grounded in facts about the actual configuration of metaphysical fundamentalia you accept (or are open to), and how the truth of propositions you think could be true in some metaphysically possible scenarios could be grounded in facts about what the metaphysical fundamentalia would be like in those scenarios.

Obviously, we don't expect the Nominalist to be able to say what fundamentally grounds various physical magnitude facts, since we don't yet know the true fundamental laws of physics and don't know which magnitudes are fundamental. But if no imaginable story about grounding could be told by the Nominalist while the Platonist could tell many such stories, this would significantly cut against Nominalism.

11.4.2   Sideran Framework

For clarity in talking and thinking about grounding, I will use the following basic framework taken from Sider (2011). However, I don't think much I say will depend on this particular choice of framework. There are three elements to consider.

First, we have a concept of **fundamentality**, which Sider identifies with joint-carvingness (in the sense in which the predicate "is an electron" is intuitively more joint-carving than the notion "is an electron or a cow"). Importantly, this question of joint-carvingness is not just supposed to apply to predicates but also to all other elements of our ideology, including variant existential and universal quantifier meanings. Notions can be more or less fundamental, and a notion qualifies as fundamental simpliciter if it is maximally fundamental. In particular, Sider takes there to be a single maximally fundamental existential quantifier sense. And fundamental objects are objects that exist in this unique maximally fundamental quantifier sense.

Second, Sider endorses the following principles which connect the idea of fundamentality qua maximal joint-carvingness to expectations about some truths grounding/explaining all other truths:

- "Completeness: Every non fundamental truth holds in virtue of some fundamental truth."
- "Purity: Fundamental truths involve only fundamental notions."

Third, Sider ultimately cashes out the "in virtue of" notion in terms of the existence of a metaphysical semantics which accounts for language users' behavior by systematically tying their claims/utterances to claims involving only fundamental (i.e., maximally joint-carving) notions.[14]

I will differ from Sider (2011) in understanding "Platonism" to mean accepting the existence of mathematical objects (in our current quantifier sense), not our most fundamental quantifier sense.[15] However, this is a mere terminological difference, and nothing turns on it.

Some remarks of Sider's may nicely flesh out the distinction between reference and grounding indispensability worries (Sider 2011). Specifically, Sider advocates a project of metaphysical semantics (something philosophers might do when nominalistically regimenting set theory or applied mathematics) which differs from linguistic semantics as follows. Both projects use notions like reference and try to explain why people say the things they do. However, the aims of metaphysical semantics differ from those of linguistic semantics in a few ways.

For one thing metaphysical semantics aims to illuminate relationships between what people say and fundamentalia, while linguistic semantics does not. Sider writes, "Metaphysical semantics is more ambitious [than linguistic semantics] in that by giving meanings in fundamental terms, it seeks to ... show how what we say fits into fundamental reality." Additionally, metaphysical semanticists don't attempt to assign meanings in a way that matches facts about sentences' syntactic form or illuminates what can be rationally derived from them a priori, or what can be known by conceptual competence alone as linguistic semanticists often do. As Sider (2011) puts it:[16]

[A person doing metaphysical semantics] is ... not trying to integrate her semantics with syntactic theory ... And she is free to assign semantic values that competent speakers would be incapable of recognizing as such, for she is not trying to explain what a competent speaker knows when she understands her language. She might, for example, assign to an ordinary sentence about ordinary macroscopic objects a meaning that makes reference to the fundamental physical states of subatomic particles. And she might simply ignore Frege's ... puzzle of the cognitive nonequivalence of co-referring proper names, since she is not trying to integrate her semantics with theories of action and rationality.

---

[14] Technically, appeal to the metaphysical semantics lets Sider eliminate the "in virtue of" notion from his theory and restate completeness as follows: "New completeness: Every sentence that contains expressions that do not carve at the joints has a metaphysical semantics." Sider's examples of such a metaphysical semantics often have the form of a truth theory "Sentence S of L is true in L iff ϕ" (where ϕ is a sentence involving only fundamentalia). But he writes "Metaphysical semantics are not required by definition to take any particular form. They must presumably be compositional in some sense (since they must be explanatory and hence cast in reasonably joint-carving terms, and must contend with infinitely many sentences). But this still allows considerable variation" (Sider 2011).

[15] Note that this difference in terminology doesn't reflect a commitment by Sider to only use the most fundamental quantifier sense when doing philosophy or even metaphysics. He also accepts that one sometimes does philosophy using less fundamental quantifier senses. It is merely a pure terminological difference.

[16] See Section 11.4.2 for more about grounding and the minimal Sideran framework I'll adopt in the following chapters addressed to philosophers who embrace the projects of traditional metaphysics.

11.4.3   Morals

I will say much more about the Reference and Grounding worries in Chapter 14, when discussing a case where Quinean and Explanatory indispensability worries look like they might be uncontroversially solvable, but Grounding and Reference worries pose a serious challenge (for some Nominalists).

For now, however, I just want to note three things. First (much as has already been pointed out in the case of Explanatory indispensability worries), Grounding and Reference worries remain even if we dismiss the need to literally state our best theory, as easy road Nominalists do.

Second, nominalists shouldn't be expected to provide a single account which simultaneously answers the Reference and Grounding indispensability worries. For, a paraphrase strategy will plausibly do a better job at answering Reference worries and explaining how we can finitely learn and grasp the infinitely many different propositions we can understand if it sticks close to surface grammar (with maybe a few divergences motivated by linguistics or cognitive science). On the other hand, a paraphrase strategy which is intended to answer questions about metaphysical grounding will likely get better (more plausible, attractive and explanatory) by going far away from surface grammar and explaining how heterogeneous facts can be cunningly grounded in facts involving a tiny fundamental ontology and ideology.

Third, I take the dialectical point about Quinean indispensability worries made in Section 2.3.2 to apply to Explanatory, Grounding and Reference worries as well. That is, I take it that addressing these worries only requires dispelling the impression that *rejecting the existence of mathematical objects* prevents you from doing something, e.g., telling a satisfactory story about our ability to explain certain things, finitely learn a language which lets us draw certain distinctions. Sometimes unrelated philosophical puzzles make it hard to tell an attractive explicit story about reference and grounding *whatever* you say about mathematical objects (i.e., for reasons that apply equally to the Platonist). Failure to provide an attractive analysis in these cases shouldn't count against the Nominalist.

12 Modal If-Thenist Paraphrase Strategy

With this picture of indispensability worries in mind, let's turn to the question of when the Nominalist can answer them.

In this chapter I will introduce a key tool in the arsenal of the Nominalist of Chapter 10: a general Nominalist paraphrase strategy for replacing claims about mathematical objects with claims about logical possibility.

This paraphrase strategy follows Hellman (1994) in putting a modal twist on familiar if-thenism, but is developed using the conditional logical possibility operator rather than Hellman's machinery. Roughly speaking, the idea will be that our nominalistic translation $T(\phi)$ of the Platonist's sentence ϕ says: it's logically necessary, fixing the facts about all relevant non-mathematical structures, that if there were also mathematical structures then ϕ.

I'll show how this nominalization strategy can be applied to any Platonist sentence ϕ satisfying a certain "definable supervenience" condition. Then I'll note that, where defined, the Nominalist paraphrases provided by this strategy will let us answer classic Quinean Indispensability arguments in the sense specified in Section 11.2. That is, it will let us transform a Platonist theory into a nominalistic theory which – the Platonist must acknowledge – matches the intended (possible worlds) truth conditions for that Platonist theory.

One might worry that the if-thenist form of these paraphrases makes them objectionably instrumentalist and unexplanatory. But in Section 13.6, I'll argue that this is not the case. In fact, in certain central cases, we'll see that relevant Nominalist regimentations of scientific theories are plausibly *explanatorily better* (more general, powerful and illuminating) than corresponding versions of the same theories. The basic structure of this paraphrase strategy will also be useful to help explicate and develop a general neo-Carnapian philosophy of language and a more realist approach to mathematical objects outside set theory (as we will see in Chapter 15).

12.1 Modal If-Thenist Paraphrase Strategy

12.1.1 Motivating Example

To motivate and begin to concretely explain the modal if-thenist nominalization strategy, consider the following sentence:

CRITICS: Some critics only admire each other.

A Platonist who believes in sets of critics could Platonistically formulate CRITICS as follows:

$CRITICS_P$: There is a set-of-critics x such that, for all y and z, if $y \in x$ and y admires z then $z \in x$.

Now our modal if-thenist paraphrase strategy lets us capture this claim as $T(CRITICS_P)$.

$T(CRITICS_P)$: $\Box_{\text{critic, admires}}$ [If there are (objects with the intended structure of) a single layer of sets-of-critics under elementhood, then (it's true in this structure that) there's a set-of-critics x such that, for all y and z, if $y \in x$ and y admires z then $z \in x$.][1]

This says (roughly) that necessarily if the actual structure of critics and admiration were supplemented with extra objects with the structure the Platonist takes the sets-of-critics to have, then the Platonist's claim $CRITICS_P$ would be true.

Intuitively (from a Platonist point of view) this claim has the same truth conditions as the original claim. The truth-value of $CRITICS_P$ is completely determined by the structure of how critic, admiration, set-of-critics apply.

We can say there are objects with the intended structure of the sets-of-critics under elementhood (when considered under some otherwise unused relations S and E) by conjoining the following:

- The claim that there are sets corresponding to "all possible ways of choosing" some critics: $\Box_{\text{critic}, S, E}$ (there's a set which contains exactly the critics who are happy).[2]

  Intuitively this captures the appeal to all possible ways of choosing by saying that it's logically necessary (fixing the structure of the critics, sets-of-critics and elementhood) that however "happy" applies there will be a set-of-critics which contained exactly the happy critics.
- A collection of first-order conditions that are easy to formulate, e.g., claims that the sets of critics are extensional, and that sets-of-critics only have critics as elements.

Call the above conjunction D (I'll later call this the Definable Supervenience condition). Then we have the following translation:

$T(CRITICS_P)$: $\Box_{\text{critic,admires}}[D \rightarrow CRITICS]$

12.1.2 Definitions

With this motivating example in mind, I will now explain the modal if-thenist translation strategy. This comes in two parts.

---

[1] Note that I use the terms set and $\in$ for readability purposes only. Any sentence produced by uniformly substituting a predicate P and a two-place relation R (without collision) for "set-of-critics" and "$\in$" will work equally well.

[2] C.f. the tools we used to describe a full-width layer of sets in Section 11.2.

First, there's a definable supervenience condition, which specifies the intended structure of all the "extra" objects and relations the Platonist believes in, in terms of their relationship to objects and relations the Platonist and Nominalist can agree on.[3] Second, there's a modal if-thenist framework which we plug this definable super-venience description into.

To explain both elements above, let me start by introducing some definitions. One might try to define nominalistic vocabulary as vocabulary which, with metaphysical necessity, applies only to non-mathematical objects. However, even the predicate for "real numbers" would satisfy that definition if Nominalism is true (for in this case, it's metaphysically necessary that nothing is a real number). So, instead, we use the following definition.

Definition 12.1 (Nominalistic Vocabulary). A predicate or relation R counts as **nominalistic** vocabulary iff the Platonist accepts that it is metaphysically necessary that the extension of R contains only objects that the Nominalist would admit exist. For example, "is a cat" is a nominalistic predicate and "is taller than" is a nominalistic relation.

Definition 12.2 (Platonistic Vocabulary). A predicate or relation is **Platonistic** iff it is not nominalistic.

Thus, Platonistic vocabulary includes not only pure mathematical vocabulary[4] but also applied mathematical vocabulary[5] and relations which (the Platonist thinks) relate mathematical objects to non-mathematical objects.[6]

Now let's turn to the definable supervenience condition. Intuitively, speaking, the **definable supervenience** condition says a description D uniquely describes the mathematical structures the Platonist accepts, using only nominalistic facts. At each metaphysically possible world, D uniquely "pins down" the pure and applied mathematical structures the Platonist believes in (given the facts about non-mathematical structures that Platonists and Nominalists agree on at that world).

We can specify what it takes for a sentence D to be a definable supervenience condition for a Platonist language formally, by generalizing the notion of categoricity to a concept of certain descriptions of a structure being "categorical over" the facts about a certain part of that structure.

The idea here is that (just as we can completely specify the structure the Platonist takes the natural numbers to have using logical vocabulary alone), we can completely specify the intended structure of the goats and sets of goats, using only facts about the goats and logical vocabulary. We believe certain things (expressible using the conditional logical possibility operator) about what the relationship between the goats and the sets of goats is supposed to be like, such that, for any way of fixing the goats

---

[3] The latter terms whose extensions the Platonist and Nominalist can agree on might include relations like "dog," "bites" and "is more massive than," but not "number" "plus" or "the mass of . . . in grams is . . . ."

[4] Such as the predicate "Is a number" or the relation +.

[5] Such as the predicates "Is a set of goats," or "is a function from the goats to numbers."

[6] ". . . has more than . . . fleas" and "is an element of" and ". . . is a function from cats to numbers which maps . . . to . . . ."

structure there's only one way that the overall goats-and-sets-of-goats structure could be (which would make our beliefs true).

So a description $D(N_1, \ldots, N_m, P_1, \ldots, P_n)$ is categorical for the $P_1, \ldots, P_n$ over $N_1, \ldots, N_m$ if the facts about how $N_1, \ldots, N_m$ apply completely determine how $P_1, \ldots, P_n$ apply – and indeed the whole $N_1, \ldots, N_m, P_1, \ldots, P_n$ structure[7] given that D is true.

We can define this notion using the conditional possibility operator, and our definition of isomorphism (Definition 7.4).

Definition 12.3 (Categorical Over). D is a categorical description of the relations $\mathcal{P} = P_1, \ldots, P_n$ over $\mathcal{N} = N_1, \ldots, N_m$ (where $\mathcal{P} \cap \mathcal{N} = \phi$) just if $(D[N_1, \ldots, N_m, P_1, \ldots, P_n] \wedge D[N_1, \ldots, N_m, P_1/P'_1, \ldots, P_n/P'_n] \rightarrow \mathcal{N} \cup \mathcal{P} \cong \mathcal{N} \cup \mathcal{P}')$

Using this we can now define the definable supervenience condition, i.e., the condition we expect our Platonist paraphrase to satisfy.

Definition 12.4 (Definable Supervenience Condition). A sentence D is a definable supervenience condition, specifying how the application of some Platonistic vocabulary P definably supervenes on that of some nominalistic vocabulary $\mathcal{N}$ if and only if the following conditions hold:

- (from a Platonist POV) D is metaphysically necessary;
- $\Diamond_\mathcal{N} D$, i.e., the Platonist isn't supposing the existence of incoherent objects, and indeed it's logically necessary that the $\mathcal{N}$ structure can be supplemented with Platonistic structure in the way that D requires;
- D is content-restricted to $\mathcal{P}, \mathcal{N}$;
- D is a categorical description of the $\mathcal{P}, \mathcal{N}$ structure over the $\mathcal{N}$ structure.

Note that all the pure and applied mathematical structures commonly used in applied mathematics (reals, complex numbers, classes of physical objects, functions from physical objects to mathematical objects etc.) can be straightforwardly given such a definable supervenience condition.

When we have a suitable definable supervenience condition D, we can translate every sentence ϕ which is content-restricted to the total list of relations in the Platonist's language as follows:

$$T(\phi) = \Box_\mathcal{N} (D \rightarrow \phi)$$

Intuitively, this says that it's logically necessary, given the structure of objects satisfying the list of nominalistic relations $\mathcal{N}$, that *if* there were (objects with the intended structure of) relevant mathematical objects then ϕ would be true. Note that the Platonist must believe it is always logically possible to supplement the actual objects with

---

[7] So, for example, if the sets of people, along with set membership, $(S_{people}, \in_{people})$ is categorical over the people P it's not just true that the number of sets of people is totally determined by what people exist but also facts such as whether or not any set of people is a person must also be determined.

objects that behave like the platonic objects and satisfy D, because they think such objects exist.

12.2   A More Detailed Example

To clarify how this strategy can be applied to more complex cases, consider a Platonist who believes in three types of mathematical objects: natural numbers, sets of goats and partial functions from goats to natural numbers.

Consider the following sentence:

GOATS: There are a prime number of goats.

The Platonist will formalize this statement with a sentence like the following:

GOATS: There's a $1 - 1$ function[8] f, such that f maps the goats onto an initial segment of the natural numbers, from 0 up to, but not including, some prime number n.

Can we nominalize this sentence? Yes. Our first step is to note that GOATS is implicitly content-restricted to a certain list of relations: natural number, set of goats, etc. It doesn't involve unrestricted quantification, and its truth-value must be the same in any logically possible scenarios which agree on this structure. Now, can we write down a definable supervenience sentence D (call it D[numbers, goats-to-numbers functions]) which categorically specifies how all the relations on this list apply in terms of how the Nominalist relations on the list apply? We can write such a D by conjoining the following:

- A categorical description of the natural numbers $\text{PA}_\diamond$ (i.e., a sentence which uniquely pins down how the Platonist thinks $\mathbb{N}, S, +, \times$ apply, up to isomorphism).
- A sentence which pins down the structure of "all possible" partial functions from goats to numbers,[9] given the structure of the goats (and numbers).
- A collection of "Julius Caesar sentences," i.e., sentences specifying how the mathematical objects are supposed to relate to the non-mathematical objects. For example, we might say that the numbers are supposed to be distinct from the sets of goats, functions from goats to numbers, etc.[10]

Note that there are only two things that can't be obviously formulated in first-order logic in my description of the supervenience description D[numbers, goats-to-numbers-functions] above: the categorical description of the natural numbers, and the description of the partial functions from goats to numbers.

Recall that we saw how to categorically describe the natural numbers with a sentence $\text{PA}_\diamond$ in Section 4.3.2.1. What about describing the structure of partial functions from the goats to the numbers? We can nominalistically formalize this in

---

[8] Here I treat functions as just another kind of mathematical object.
[9] I will treat these as free-standing mathematical objects.
[10] This may include specifying that the numbers and sets of goats are distinct from all (the finitely) many types of non-mathematical objects relevant to the physical theory to be translated.

the same way. Assume the Platonist's language has relations "function()" and "maps()" such that maps(f, x, y) iff f is a function that maps x to y, i.e., $f(x) = y$. We can informally pin down the structure we want by saying two things:

- There are functions witnessing all possible ways of mapping some of the goats to some of the numbers.[11]
- There are no more functions than needed to ensure this (i.e., every function maps only goats to numbers[12] and the functions are extensional).

The second claim is easy to formalize in FOL. And we can write the first using second-order relation quantification as follows:[13]

$\forall R$[If R is functional and only relates goats to numbers then $(\exists x)(\text{function}(x) \wedge (\forall y)(\forall x)$ $[\text{maps}(x, y, z) \leftrightarrow R(y, z)).]$

We can rewrite this in the language of logical possibility, using any two-place relation that doesn't figure in the body of scientific theorizing we want to translate. For example, I will pick "eucratises":[14]

$\square_{N, \text{function,maps,goat}}$[If eucratises applies functionally and only relates goats to numbers then $(\exists x)(\text{function}(x) \wedge (\forall y)(\forall x)$ $[\text{maps}(x, y, z) \leftrightarrow \text{eucratises}(y, z)).]^{15}$

It's logically necessary given the structure of the goats, numbers and functions from goats to numbers, that if eucrastises only relates goats to numbers and applies functionally there's a function x that relates goats to numbers in the same way.

Given D[numbers, goats-to-numbers functions] the Nominalist can translate the Platonist's formalization of the claim that there are a prime number of goats into a nominalistic version of this claim, $T(\text{GOATS})$, as follows:

$T(\text{GOATS})$: $\square_{\text{goat}}(D[\text{numbers, goats-to-numbers-functions}] \rightarrow \phi_{\text{GOATS}})$

Intuitively, this says that it's logically necessary, given the structure of the goats, that *if* there were (objects with the intended structure of) the numbers, sets and functions from goats to numbers then GOATS would be true.

Furthermore, we can show that the Platonist must agree that this translation is true at the correct set of metaphysically possible worlds (i.e., the worlds at which *they* take ϕ to be true); they must think that it's metaphysically necessary that $T(\phi) \leftrightarrow \phi$.

At each possible world w, the truth-value of GOATS is completely determined by the structure of goats, functions and numbers at that world.[16] And (according to the Platonist) the latter structure is completely determined by the structure of the goats at w

---

[11] Or in the limiting case of the partial function that's not defined anywhere, pairing no goats with numbers.

[12] That is, $(\forall x)(\forall y)f(x) = y \rightarrow \text{goat}(x) \wedge \text{number}(y)])$.

[13] For every relation R that only relates goats to numbers (in that $(\forall x)(\forall y)Rxy \rightarrow x$ is a goat and y is a number] which is functional (R is functional iff $(\forall x)(\forall y)(\forall z)[(Rxy \wedge Rxz) \rightarrow y = z])$ corresponds to a function f.

[14] This is the relation x and y stand in when x restores y to the correct balance of humors (eucrasia).

[15] That is, $\square_{N, \text{function,maps,goat}}$ [if nothing eucratises two distinct things and only goats eucratise and only numbers get eucratised then $(\exists x)$ (function$(x) \wedge (\forall y)(\forall x)[\text{maps}(x, y, z) \leftrightarrow \text{eucratises}(y, z))]$.

[16] Note that GOATS can be written with quantifiers restricted to objects which at least one of the Platonistic or nominalistic relations just mentioned (e.g., "goat," "set," "... is an element of ..." "number," "function," "... is a function that assigns ... to ...") apply to.

together with our definable supervenience description D. D completely pins down what sets and functions (the Platonist thinks) there are at w, given the facts about nominalistic stuff at w. There's only one logically possible way (structurally speaking) to supplement the pattern of goats at w with numbers and functions as required by the claim D, which the Platonist takes to be a metaphysically necessary truth. So GOATS is true at w if and only if it's logically necessary, given the facts about the goats at w and D, that ϕ.

So our total translation will have the following form:[17]

$$\Box_{\text{goat}}[\psi_1 \wedge \Box_{N,S}(\psi_2) \wedge \Box_{N,\text{function,maps,goat}}(\psi_3) \rightarrow \phi_{\text{GOATS}}]$$

As in the previous case, the Platonist must say that this statement is true at exactly the same metaphysically possible worlds where GOATS is true.

12.3   Clarifications and Advantages

12.3.1   Harmlessness of Platonist Science

We can show that the nominalistic paraphrase strategy produced by our translation strategy T preserves the desired inferential role of scientific sentences. It captures both inferences from applied mathematical sentences to other applied mathematical sentences, and inferences between applied mathematical sentences and observational sentences.[18]

But we can also show (see Section D.1 in Appendix D) that where we know it's metaphysically necessary that $\Diamond_N D$ (something my Platonist and Nominalist alike take themselves to know[19]) we have:

---

[17] Here ψ_1 is the part of our descriptions of the numbers and functions from goats to numbers which is straightforwardly stateable in FOL.

[18] At least, this claim holds on the plausible assumption that the former can be understood as content-restricted to some Platonist vocabulary and the latter can be understood as content-restricted to some Nominalist vocabulary.

    I tentatively hypothesize that no non-negotiable scientific practice requires unrestricted quantification. To roughly motivate this idea, we might say that when concerned with physics or biology, scientists don't (and needn't) concern themselves with talking about what fictional characters, or marriage licenses, could be like. So, we shouldn't need to use sentences whose quantifiers are restricted to range over literally all objects, including these irrelevant ones. It suffices to use quantifiers which range over, e.g., all (relevant applied mathematical objects and) physical particles and spatial points, rather than statements of universal quantification.

    Also note that formulating all our physics room talk with restricted quantifiers doesn't require us to have any illuminating conception of each of the types of physical objects we are quantifying over, like "quark" or "boson" (thanks to Vann McGee for pressing this point in conversation). We can take our quantifiers to range over objects satisfying some rather broad uninformative notion like "fundamental physical object" or physical object. Or, if a Leibnitizan regress appears to exist, that the physics relevant particles/particles at a certain level or lower are such and such. This is important because it means that we can ask questions like "How many different kinds of fundamental physical particles are there?" without having much of a sense of what these particles are like.

[19] We will generally be able to derive this from the fact that $\Box_N D$, i.e., that it's *logically* necessary that however the nominalistic relations apply it's possible, holding fixed those nominalistic relations, that D.

Theorem 12.1 *Suppose that Φ, Ψ are content-restricted to $\mathcal{P} \cup \mathcal{N}$ and $\vdash \Phi \rightarrow \Psi$ then $\vdash T(\Phi) \rightarrow T(\Psi)$. Furthermore if $\vdash T(\Phi) \rightarrow T(\Psi)$ then $\vdash (D \wedge \Phi) \rightarrow \Psi$.*

That is, Platonist scientific arguments from $(D \wedge \Phi)$ to Ψ (where the latter don't involve unrestricted quantification) can be easily transformed into Nominalist scientific arguments, from $T(\phi)$ to $T(\phi)$ and vice versa. Note that, for any statement v that's content-restricted to Nominalist vocabulary, $T(v) \leftrightarrow v$ is easily derivable.

Putting this together, we get that whenever a Platonist can use their Platonist assumption D conjoined with sentences $\rho_1, \ldots, \rho_n$ content-restricted to nominalistic stuff to prove ρ' content-restricted to nominalistic stuff, there is a good argument from $\rho_1, \ldots, \rho_n$ to ρ'.

12.3.2   Conditional Logical Possibility and Field's Conservativity

Let me end with two points of comparison between this strategy and that famously advocated by Hartry Field in *Science Without Numbers* (Field 1980).

First, using the logical possibility operator and axioms I've proposed is helpful to those who would follow Field's paraphrase strategy as well. In this way we can cash out an intuitively appealing "conservativity" argument made by Field (1980) to account for the goodness of Platonist science from a Nominalist point of view, while avoiding worries about circularity which I will now explain.

Field (1980) wants to explain why using mathematics in the sciences is harmless and indeed helpful, despite the fact that (as he wanted to say at the time) existence claims about mathematical objects are false. He wants to say this is true because mathematical axioms are conservative (in the sense below). Reasoning with these axioms just speeds up proofs; it doesn't let us prove anything new about non-mathematical objects.

if B is any sentence, B^* is the result of restricting B to non-mathematical entities, and $M_1, \ldots, M_n$ are the axioms of a mathematical theory $M_1, \ldots, M_n$, the conservativeness of M can be expressed by the following schema:
   (C) If $\lozenge B$, then $\lozenge (B^* \wedge M_1 \wedge \ldots \wedge M_n)$. (Field 1980)

Field argues, working in ZFC, that one can always take a model of just the non-mathematical entities recognized by a theory and produce a model which also recognizes a hierarchy of sets taking those objects as ur-elements. He is criticized in Bueno (2020) for circularly using set theory to justify the claim that assuming set-theoretic axioms won't let you prove anything false about non-mathematical objects in this way.

But if we accept the notion of logical possibility and the axioms I've proposed for it, we can justify a version of Field's desired conservativity result (for suitable axioms describing mathematical objects, like a hierarchy of sets V_α up to some suitably definable height V_α) from modal principles that don't assert the existence of mathematical objects.[20]

---

[20] Let M be supervenience description for a hierarchy of sets with ur-elements chosen from the physical objects with height ω. Let $R_1, \ldots, R_n$ be some list of nominalistic vocabulary such that it's

Second, my paraphrase strategy always produces finitely stateable theories where it applies, and in Chapter 14 I'll argue that it can be applied to solve the physical magnitude problems (at least for purposes of Quinean indispensability, if not reference and grounding) which drove Field to appeal to infinitely many different sentences satisfying a schema, rather than producing a single sentence that formalizes the scientific theory at issue.[21]

metaphysically necessary that: only non-mathematical objects are related by these relations and every non-mathematical object has at least one of these relations apply to it.

Then for any list of nominalistic relations $R_1, \ldots, R_n$, we can use axioms like those proposed in Part II to show that $\Box\Diamond_{R_1,\ldots,R_n} M$. That is, it's necessary that whatever the nominalistic $R_1, \ldots, R_n$ structure is like, this structure can be supplemented with some new objects playing the roles of sets in V_ω with ur-elements.

And given $\Box\Diamond_{R_1,\ldots,R_n} D$, we can prove the analog of Field's conservativity statement. Consider an arbitrary nominalistic sentence β, corresponding to Field's B* (a description of what's supposed to be happening with regard to the physical objects), hence content-restricted to our list of Nominalist relations $R_1, \ldots, R_n$ and satisfying $\Diamond\beta$.

We can prove that if $\Box\Diamond_{R_1,\ldots,R_n} M$ and $\Diamond\beta$, then $\Diamond(\beta \wedge M_1 \wedge \ldots \wedge M_n)$, as follows. Assume that $\Box\Diamond_{R_1,\ldots,R_n} M$ and $\Diamond\beta$. Enter this $\Diamond$ context. Then we know β. We can import our assumption that $\Box\Diamond_{R_1,\ldots,R_n} M$, as it is content-restricted to the empty list of relations. Then we can infer $\Diamond_{R_1,\ldots,R_n} M$ by $\Box$ Elimination (Lemma B.4 of Section B of the online appendix). Now because β is content-restricted to $R_1, \ldots, R_n$ we can infer that $\Diamond_{R_1,\ldots,R_n}(M \wedge \beta)$ by Axiom 8.6. So leaving the diamond context we have $\Diamond\Diamond_{R_1,\ldots,R_n}(M \wedge \beta)$, hence $\Diamond_{R_1,\ldots,R_n}(M \wedge \beta)$ by Axiom 8.2 (Diamond Elimination) and $\Diamond(M \wedge \beta)$ by Axiom 8.3 (Diamond Ignoring).

Thus we can use modal nominalistic reasoning about logical possibility, rather than set theory (as Field does) to show that assuming the existence of a hierarchy of sets with ur-elements over the physical objects you are currently talking in terms of up to V_ω is harmless, and doesn't let you prove anything false or unjustified about these non-mathematical objects.

[21] However, certain disadvantages may also be admitted. Most obviously, accepting the conditional logical possibility operator is controversial (though, recall, Field himself advocates accepting a primitive logical possibility operator and uses it in his argument for conservatism). Also, the kind of paraphrases of physical magnitude statements provided will not be as attractively "intrinsic" in the way Field wants.

13 Explanatory Indispensability

13.1 Introduction

With the strategy outlined in Chapter 12 for nominalistically paraphrasing Platonist scientific theories in mind, let's now turn to the Explanatory Indispensability challenge.

In this chapter I will argue that translations produced by the strategy from Chapter 12 can be used to answer Explanatory Indispensability challenges as well as Quinean demands for literal statements. Much of the chapter will consider a certain prominent and representative case where mathematical objects have been claimed to be explanatorily indispensable: Baker's Magicadas explanation (Baker 2005). I'll note that we can (Platonistically) logically regiment Baker's Platonist explanation for why certain cicadas tend to have life cycles that are a prime number of years, in such a way that that the nominalization strategy from Chapter 12 can be applied.

I then will argue that the resulting nominalistic theory is explanatorily at least as good as (and arguably even better than) the original Platonist explanation. In doing this I hope to address a natural worry that (despite being true at the right set of possible worlds), the if-thenist structure of such nominalizations of scientific theories prevents them from providing good explanations. I will further argue that the nominalistic paraphrase strategy deployed here improves on existing Nominalist paraphrases strategies.

13.2 Motivating Case: Three Colorability

To illustrate how the conditional logical possibility operator is useful for providing illuminating nominalistic mathematical explanations of physical phenomena – and why one might think these explanations *improve on* Platonist ones – let's return to the case of three colorable maps.

Suppose that a certain map (perhaps one with infinitely many countries) has never actually been three colored. A good explanation for this fact might be that (in a mathematical sense) the map isn't three color.

A natural Platonist explanation along these lines goes as follows:

Platonist Non-Three-Colorability: There is no function (in the sense of a set of ordered pairs) which takes all countries on the map to numbers 1, 2 and 3, in a such a way that adjacent countries are always paired with distinct numbers.

However, we can also consider a Nominalist version of this explanation, as follows.

Modal Non-Three-Colorability: $\neg\Diamond_{\text{adjacent,country}}$ Each country is either yellow, green or blue and no two adjacent countries are the same color (and each country is exactly one color).

The above modal explanation can seem to be at least as good, indeed better than the Nominalist explanation.

In particular, one might argue that the Platonist non-three-colorability principle only intuitively explains the fact that the map is not three colored because we have background knowledge of a relationship between set-theoretic facts and the modal facts above. Specifically, we think that there are functions corresponding to *all possible ways* of pairing countries with one of the numbers 1, 2 or 3, and hence all possible ways of "choosing" how to color these countries. And if we didn't accept this, then we would have no reason to suppose that there really was a function corresponding to a potential three-coloring.[1]

Thus, it may seem that the real explanatory work here is being done by the modal principle; claims about what mathematical objects exist, only witness logical possibility facts and don't really add anything over and above such facts to the explanation.

Indeed, one might argue that the Platonist account only seems explanatory and satisfying because the modal facts (about conditional logical possibility) make us feel that we've explained the phenomenon. If we imagine giving up the assumption that there are sets/functions corresponding to all logical possibilities for how colors could apply, then the Platonist story no longer feels explanatory. We would no longer be able to infer from the fact that there's no function coding a way of three coloring the map to the conclusion that the map isn't (and couldn't be) three colored.

A Platonist might resist this above argument by saying that they get from set and function existence to the conclusion the map isn't three colorable in a different way. The Platonist might say they that this inference is justified by appealing to something like the following non-modal comprehension schema – rather than to any modal notion like conditional logical possibility.

Ur-element Comprehension Schema: For every English-definable predicate ϕ definable with parameters, which only applies to non-sets:[2]

$$(\exists x)\Big[\text{set}(x)\wedge(\forall y)(y\in x\leftrightarrow\phi(y))\Big]$$

But note that this schema only asserts that there are sets corresponding to every way that some predicates in our current language (will) *actually* apply to some objects. Thus, it doesn't capture our intuitive idea that the mere structure of how the countries are related by adjacency explains why this map will never be three colored.

---

[1] The argument has some similarities to the argument of Section 4.1.3 that we need to have a notion of logical possibility that's distinct from having a set-theoretic model, even though the completeness theorem (ultimately) winds up showing that the two notions are extensionally equivalent for first-order claims.

[2] So, assuming certain popular axioms of set theory with ur-elements like that given in McGee (1997), only applies to set-many objects.

It also doesn't explain why we should expect it to be *physically and metaphysically necessary* that (intrinsic duplicates of) this map won't be three colored.[3] And perhaps it doesn't explain why we'd expect an analog of non-three colorability to hold for all triples of properties we might introduce via some "logic preserving change to our language" that adds new predicates.[4]

Accordingly, I think considering this explanation provides a nice motivating example for how nominalistic-mathematical explanations for scientific facts can be as good (and in some senses even intuitively better than) Platonist ones. Perhaps (in this specific case) the modal formulation of our explanation even matches ordinary language better than the Platonist one. For we tend to express the above thought about maps being three color*able*, rather than ontologically about maps *having three colorings*.

13.3   Magicadas

13.3.1   An Argument for Explanatory Indispensability

Now let's turn to the main case to be considered in this chapter. *Stanford Encyclopedia* (Colyvan 2019) summarizes the Magicadas case and testifies to its prominence in the literature as follows:

One example of how mathematics might be thought to be explanatory is found in the periodic cicada case (Yoshimura 1997 and Baker 2005). North American Magicadas are found to have life cycles of 13 or 17 years. It is proposed by some biologists that there is an evolutionary advantage in having such prime-numbered life cycles. Prime-numbered life cycles mean that the Magicadas avoid competition, potential predators, and hybridization. The idea is quite simple: because prime numbers have no non-trivial factors, there are very few other life cycles that can be synchronized with a prime-numbered life cycle. The Magicadas thus have an effective avoidance strategy that, under certain conditions, will be selected for. While the explanation being advanced involves biology (e.g., evolutionary theory, theories of competition and predation), a crucial part of the explanation comes from number theory, namely, the fundamental fact about prime numbers.

Baker (2005) argues that this is a genuinely mathematical explanation of a biological fact. There are other examples of alleged mathematical explanations in the literature, but this remains the most widely discussed and is something of a poster child for mathematical explanation. (Colyvan 2019)

This description doesn't specify a precise explanation or explanandum. However, I take the following principle to be a (simple but) fairly representative example of the kind of explanation at issue. Although not many animals have multi-year hibernation cycles like cicadas, cicada predators can have regular multi-year cycles of population spikes and troughs. In this case, there will be evolutionary benefit to cicada species

---

[3] Perhaps you could add a *sui generis* law that all instances of the comprehension schema hold with metaphysical necessity. But if one accepts the notion of logical possibility then explanations that appeal to such a law (even granting that it's a genuine law) seem less direct and illuminating than explanations by appeal to general laws of logical possibility which you already accept.

[4] See McGee (1997).

avoiding spikes in predator population. And a mathematical principle like the following can help us connect these constraints to the conclusion that cicadas have (or are likely to have) a prime numbered life-cycle:

It's metaphysically/physically/mathematically necessary that, if for some n:

- premise 1: there are predator species which have population spikes of length p, for each $p \leq n$; and
- premise 2: the Magicadas' life cycle (re: emerging from hibernation every c years) is optimal with regard to minimizing overlap with predator population spikes, among a set S of "biologically viable options,"[5] which include some prime number $> n$ and are all less than $2n$,

then the Magicadas have a prime number length life cycle.

When combined with the empirically motivated claim that premise 1 is satisfied (with regard to some specific natural number n) and premise 2 is satisfied (with regard to the same number n and the set S of numbers satisfying some number theoretic predicate ϕ), this claim will entail that Magicadas have a prime number length life cycle. And we can see that this principle is a mathematical truth by noting that any composite number relatively prime to every predator cycle length $p < n$ must be equal to or greater than $2n$. Thus, all possible cicada life cycle lengths which are both relatively prime to all these predator cycles (hence minimizing overlap with them) and less than $2n$ will be prime.

Thus we have a mathematical principle which plays a key role in explaining a physical phenomenon (that the Magicadas have a prime number length life cycle), and we want to know whether the existence of numbers is necessary to that work.

Note that I wrote the above principle as a modal claim "It's metaphysically/ physically/mathematically necessary that," because I take the fact that the conditional claim about life cycles is a law to do explanatory work. If we just believed that this material conditional happened to actually be true in our world (and didn't see why it mathematically had to be) it wouldn't feel like a good explanation.

13.3.2   Existing Nominalizations

In existing work, Rizza (2011) argues (correctly I think) that Baker's story about Magicadas presents a genuinely mathematical explanation for a scientific phenomenon, but not one that commits us to the existence of mathematical objects. He backs this up by providing a particular nominalistic paraphrase for Baker's Platonist mathematical explanation of the Magicada phenomenon.

---

[5] That is, for each predator type, the long-run fraction of times Magicadas with this life cycle would overlap with predator population spikes – given that they overlap at least once – is less than or equal to the long-run fraction of times Magicadas with any of these alternative biologically viable option life cycles would overlap. I take the argument to be presuming (plausibly enough) that because predators evolve too you presumably can't avoid predation by, e.g., being around on alternate years w.r.t. some predator which also has a two-year cycle.

Rizza (essentially) points out that we can reconstruct a version of the Platonist account which only quantifies over some initial segment of the natural numbers. He then proposes to replace quantification over numbers with quantification over time points in some evenly spaced sequence of years with a starting point. Using relations like congruence between temporal intervals "there's as much time between a and b as between c and d," we can then define an analog to successor, plus, times, etc. on this sequence of years, creating a temporal structure (a sequence of points in time) that's isomorphic to some initial segment of the natural numbers. Thus, we can reformulate the Platonist argument that (under relevant assumptions) we should expect to see cicadas with prime length life cycles by systematically replacing claims about this initial segment of the numbers (and mathematical relations on it) with corresponding claims about this initial segment of the years.

Now in order to state the explanans and explanandum in Baker's Magicadas explanation nominalistically, we need to somehow make claims about cicada life cycles, and the biologically viable options for alternative cicada and predator life cycles. So, a Platonist logical regimentation of the explanation might use the following relations:

- "species . . . has a life cycle of length . . ." between animal/species or populations of cicadas and numbers;
- "the biologically viable options for cicada/predator life cycles (the life cycles cicadas/predators could have if selection favored it) are exactly the numbers within the range . . . to . . . years."

Rizza's nominalistic paraphrases use analogous relations between animals/species and points in the finite sequence of temporal points. (To make this feel natural, we might think of the first relation as meaning something like "x has a life cycle with length such that if x emerged during the year designated then it would next be disposed to awake in year y, and then to repeat the cycle and give birth to children who would.")

In this way, Rizza argues that we can dispense with mathematical objects in Baker's example, by giving the nominalistic mathematical explanation from Section 13.3.1 instead.

Notably, Rizza's nominalization strategy resembles and takes inspiration from Field's influential strategy for nominalizing physical magnitude claims in Field (1980) (which I won't summarize). Both paraphrase strategies (in effect) assume the existence of a physical structure which resembles a mathematical structure used in the theory to be paraphrased (some temporal points isomorphic to initial segment of the natural numbers in one case, and an infinite plurality of spacetime points isomorphic to the reals in the other case). Both then appeal to measurement theoretic uniqueness theorems to show that their paraphrase delivers correct truth-values in all scenarios where the relevant physical assumption holds. One might argue Rizza's story has an advantage over Field's in requiring weaker physical assumptions, as Rizza only needs to assume there are a finite number of temporal points (but I will question whether the benefits of this are worth the cost in the next section).

13.3.3   Weaknesses of Rizza's Paraphrase Strategy

Rizza's paraphrases require finding a copy of the mathematical structures mentioned by the Platonist theory they're trying to paraphrase in the physical world. This significantly limits how widely they can be applied.

For instance, one might well want to appeal to mathematical structures too large to have physical models in attempting to most illuminatingly explain some (logically/mathematically necessary) physical phenomenon. Consider how sometimes the most illuminating proof of some fact about the real numbers involves considering them within the complex numbers. Similarly, one might expect that different and (sometimes) larger mathematical structures (e.g., segments of the hierarchy of sets) could be relevant to giving the most illuminating explanation for a mathematical phenomenon.[6] And, as Baker (2016) points out, even in cases where we can prove some science-relevant mathematical constraint on reality using relatively small mathematical structures, we can often prove a more powerful and general claim (and hence show that the law in question would hold under a wider range of cases[7]) by appealing to more varied and sometimes larger mathematical structures.

For example, one might argue that Rizza's paraphrase isn't as good an explanation as the Platonist explanation for Baker's Cicadas fact, because it's not as general. Rizza shows can state and prove that prime cicada lifecycles are required to minimize overlap for any particular value of n (and all suitably truncated versions of all needed lemmas which only talk about the initial segment). But without assuming there are an infinite number of spacetime points, he can't state (much less prove) the general theorem quoted for arbitrary values of n.

One also might worry that Rizza's paraphrase strategy doesn't let us account for the biological significance that facts about the greatest common divisor (gcd) have on the Platonic explanation. For example, the original biology paper Rizza cites uses remarks about infinite sequences like, "Note that [a certain fraction] yields an average valid for $t \Rightarrow \infty$ because the process is periodic with period $[c^*p]$" (Goles *et al.*, 2001), to argue that cicada fitness goes up as $\gcd(c,p)$ goes down. However, Rizza's paraphrase strategy can't handle such infinite sequences (though Field's can).

13.4   Nominalizing Baker's Explanation

We can avoid the weaknesses mentioned in Section 13.3 by regimenting Baker's Platonist explanation with the paraphrase strategy of Chapter 12 instead.

To apply this strategy to Baker's Magicadas explanation, we need to show that one can Platonistically formalize the latter theory in a way that satisfies the definable supervenience condition from Section 12.1. So, we need a description that pins down

---

[6]   As Feferman (n.d.) notes, appeal to the existence (or at least logical possibility/coherence) of very large mathematical structures may provide our only reason for thinking that certain mathematical axioms, and hence figure indispensably in our best explanation for why no proofs inscriptions of certain kinds exist.

[7]   That is, one can show that fewer physical assumptions are necessary to guarantee that the law applies.

all relevant Platonistic structures, given the facts about how some nominalistic vocabulary applies.[8]

We can nominalistically paraphrase talk of sets of temporal points, and functions from temporal points to numbers in the same way we were able to capture talk of sets of critics and functions from goats to numbers in Section 12.2. But what about the Platonistic notions used to discuss (actual and biologically viable) life cycles?

- PlatonistActualLifecycle(x, n): an animal/species x has a life cycle of length n.
- PlatonistPossibleLifecycle(x, n): it is a biologically viable option for animal/species x to have a life cycle of length n.

These relations definably supervene on nominalistic relations of essentially the kind Rizza mentions. For example, a nominalistic version of the ActualLifecycle(x, n) might relate animals/species and pairs of temporal points.

- NominalistActualLifecycle(x, a, b) iff animal/species x is disposed to hibernate for the length of time between a and b and then repeat the cycle.

Specifically, we can uniquely specify how the relation ActualLifecycle (that the Platonist uses) behaves in terms of NomalistActualLifecycle, plus Platonist vocabulary concerning numbers and functions from numbers to years (which we've already shown satisfies the definably supervenience condition) plus a notion of temporal congruence[9] and temporal ordering, i.e., using the relations:

- TempCong(x, y, z, w): as much time passes from x to y as from z to w
- Before(x, y): temporal point x is before temporal point y.

Note that, just as we described the structure of goats, sets of goats and functions from numbers to using the techniques for mimicking second-order quantification mentioned in Section 10.1, we can categorically describe the natural number structure and uniquely pin down the intended structure of functions from these numbers to temporal points. We can then specify how the Platonist actual life cycle relation relates these "numbers" to temporal points as follows.

An animal/species x bears the Platonistic "actual life cycle length" relation to a natural number n iff x bears the *nominalistic* "actual life cycle length" relation to a pair of temporal points a, b and there are n years between a and b. And (given the truth of the definable supervenience conditions for functions from numbers to years), this will be true if and only if some function counts off n temporal points separated by 1-year intervals with $f(0) = a$ and $f(n) = b$. So, we have the following:

PlatonistActualLifecycle(x, n) iff the usual definable supervenience conditions for the numbers and years is satisfied and there are temporal points a b such that NominalistActualLifecycle(x, a, b), and there's a function f which maps the numbers from 1 to n to temporal points in such a way that

---

[8] I take it that the Platonistic paraphrase I'll propose clearly satisfies the requirement of being writable in an appropriately content-restricted fashion.

[9] This holds assuming that, like Rizza, we have some definite description of a pair of temporal points picking out a canonical year.

$f(0) = a$ and $f(n) = b$ and, for each number k, $f(k)$ is before $f(k+1)$, and the time between $f(k)$ and $f(k+1)$ is congruent to that between the beginning and endpoints of the canonical year.

13.5   Advantages and Applicability

We can immediately see how adopting this strategy removes the limitations for Rizza's strategy noted in Section 13.3. Rizza's strategy couldn't mirror Platonist theories and explanations involving very large mathematical structures because it, in effect, depended on finding a copy of all mathematical structures employed by the Platonistic theory/explanation in the physical world. Thus, it couldn't translate Platonist theories quantifying over mathematical structures too large to have models in actual space and time. This raised doubts about the applicability of this type of paraphrase strategy, and its explanatory goodness (in comparison to Platonist alternatives) where it can be applied.

In contrast, my preferred paraphrase strategy has no problem applying to Platonist theories that quantify over arbitrarily large mathematical structures (provided we have a suitable description of them). For it is logically possible that existing physical structures exist alongside arbitrarily large mathematical structures.

Thus, for example, unlike Rizza (2011), we have no problem saying that *for all* natural numbers L, if the biologically viable options for predator life cycles are those natural numbers p such that $2 \le p \le L/2$, and those for cicada life cycles are exactly those natural numbers c such that $2 + L/2 \le c \le L$, and Magicadas have life cycles favored by the type of selection for fitness discussed at the start of this chapter, they have life cycles lasting prime numbers of years.

Adopting this paraphrase strategy may also let us address some concerns which Colyvan (2001) raises about the explanatory virtues of Field's paraphrases. Colyvan suggests that Platonist formulations of physical laws provide theoretical unification by letting us articulate the idea that two very different physical systems (say, a wave in water and an electromagnetic wave) have a similar physical *structure* and obey the same differential equation.

I take the point to be that a Platonist would say that both a wave in water and an electromagnetic wave can be described by a function which satisfies the same pure mathematical description (a certain differential equation). In this case, my Nominalist can say something similar: that it is logically possible for each physical structure to exist alongside a function capturing the relevant features of the physical system, and logically necessary that a certain shared description (in this case the differential equation[10]) would be satisfied.

For example, in the case of a water wave, the Platonist would identify a function describing how the water's height at each location varies with time (and say this

---

[10] If you employ the strategy for removing mathematical vocabulary suggested at the end of the previous section, this description might be the result of uniformly substituting some other predicates/relations for mathematical predicates/relations in the Platonist's description.

function satisfies a certain differential equation). And my Nominalist would say that it's physically necessary that the physical relations that constitute the definable supervenience base for claims about these abstracta apply so that it's logically necessary (given how these relations apply) that any function which captures the height of the water as a function of time (as specified in the relevant definable supervenience condition[11]) obeys that same differential equation. Thus, both Platonist and Nominalist regimentations of our theories will make clear that the same function describes the behavior of the water wave and the electromagnetic wave.

Similar considerations address another concern Colyvan (2001) raises in the same chapter: that the Platonist can say what's correct about physical theories which get some mathematical equation right but incorrectly describe the underlying physical structures governed by that equation in some other way, and the Nominalist can't. We can also use my nominalistic paraphrase strategy to say that two physical systems "have shared structure" in the sense of being isomorphic (when considered under certain relations).

13.6   A Worry about Instrumentalism

Now let's turn to the topic of explanation. I've argued that the Nominalist version of Baker's Platonist explanation is adequate in the sense of Section 11.3.2 (so that, e.g., it imposes exactly the intended constraints on non-mathematical reality). But is it a good explanation?

One might worry that appeal to my nominalization of the Magicadas Conditional given in this chapter isn't as explanatorily useful as the Platonist original, because it has an unappealingly instrumentalist form. It might seem to resemble intuitively unattractive instrumentalist reformulations of scientific theories to avoid ontological commitment. For example, consider a version of our best actual physical theory (or Newton's), which says there is no moon and eliminates explanatory appeals to the moon, by positing suitably changed instrumentalist laws of gravitation, optics, etc. This theory doesn't assert the existence of the moon, but says (it's a law that) everything else will behave as it would *if* there were a moon with certain properties and our original physical laws applied (whether or not there is a moon).

The paraphrases associated with this moon-denying theory would be short, like mine. And the moon-denying theory makes the right predictions about everything that will happen to non-moon objects in the future. Yet there is intuitively a sense in which the actual existence of the moon does explanatory work in accounting for things like the motions of the tides, and the moon-denying theory which posits (seemingly ad hoc) variations in the laws of gravity near a certain point in the solar system seems to provide a worse explanation. Thus, although not indispensable to *stating* the constraints we expect to apply to the behavior of non-moon things, commitment to the existence of the moon very plausibly is indispensable to our best explanation of the behavior of non-moon things.

---

[11] See Chapter 5 for discussion of what this definable supervenience condition might look like.

So, a critic might wonder, how do we know that the nominalized Cicadas explanation just proposed (and all other explanations produced via the strategy outlined in Section 13.3) aren't bad in just that way? Isn't their form even suspiciously similar (both are broadly "if-thenist")?

To address this worry, I will highlight an important point of disanalogy. To explain the motion of the tides, the moon-instrumentalist needs to posit controversial alternative physical laws. They must appeal to laws which are (intuitively) inelegant and less suitable to be supported by inductive generalization than the simpler theory which says that gravity works the same everywhere. Accordingly, their overall theory strikes the Platonist as a priori less plausible than the moon endorsing theory (even if both imply all the same consequences for non-moon objects). It is certainly not something which moon advocates are already committed to accepting or take themselves to have strong independent reason to believe.

In contrast, in the Magicadas case (as we have seen), the Nominalist theory which implies all the same data about concrete objects as our Platonist theory does is not only comparably plausible but actually follows from something which the Platonist themselves *already accepts* (or has strong independent reason to accept as a law).

What about simplicity? I claim that the basic logical laws which let one derive the mathematically necessary premise in the Cicadas explanation are very simple (so far as laws go). Admittedly, Nominalist statement of other contingent physical laws produced by the paraphrase might be slightly more complex than straightforward Platonist versions of these laws. But I think with regard to these principles, the Nominalist can more plausibly say that the kind of extra complexity arrived at is harmless. They might argue that it is analogous to replacing a theory which treats heat as a fundamental property with one which reduces it to molecular motion.

The special point Baker is making with this example – that it's implausible to say that talk of mathematical objects is just a convenient fiction used to describe physical facts, because mathematical objects/facts play an explanatory role – has been answered.

13.7   Conclusion and Morals

In this chapter, I have argued that considering the paraphrase strategy of Chapter 3 strongly suggests that mathematical objects aren't indispensable to either literally stating or explaining why Magicadas have prime number length life cycles or explaining this fact. Mathematical explanations for scientific facts can often be replaced by logical explanations (and if we use the paraphrase strategy I have suggested, this needn't involve sacrifice to unificatory power).

One obvious question to ask is: how far does this generalize? I lack space to fully answer this question here. I've only considered an example. But the modal if-thenist strategy under consideration clearly generalizes to handle many extra cases. Indeed, I suspect it can be used to handle all the cases of supposed mathematical explatory indispensability that are commonly cited, *if* Chapter 14's mission of regimenting physical magnitude claims can be carried out. For, if we look at the list of purported mathematical explanations of scientific facts in Lyon (2012), the following picture emerges. The paraphrase strategy advocated in this chapter can be immediately applied to about half the cases Lyon mentions (regardless of success in regimenting physical magnitude claims). For example, interested readers will easily see how it can be used to nominalistically explain the fact that no walk satisfying the conditions for the famous Köningsburg bridge problem ever crosses each Köningsburg bridge exactly once. In the other cases, it's not clear whether this paraphrase strategy can be applied, solely because the facts to be mathematically explained involve distance and other physical magnitude functions, and it is not clear whether we can write definable supervenience conditions for these.[12]

---

[12] For example, one can explain why honeycombs have the hexagonal etc. structure that they do by saying that this structure uniquely minimizes the surface area required to contain a certain volume. That is, it's mathematically necessary that if honeycombs have a structure which has an optional volume to surface area ratio (while obeying certain structure) then they will have this particular hexagonal structure. In these cases, it seems that the mathematically necessary conditional will be a logically necessary conditional (just as in the other cases); however, because the physical antecedent and consequent involve claims about lengths and such physical magnitudes it is not clear that they can be nominalistically stated.

Easy road Nominalists might also try to use the great intuitive similarity between cases of mathematical explanation for physical facts that can be straightforwardly handled by the strategy of this chapter and those that cannot to argue that in both cases what we really have is ultimately best seen as a logical explanation of scientific facts. We just (they might say) are prevented from demonstrating this fact in some cases by the same fact that motivated people to be easy road Nominalists in the first place: that we need the fiction of mathematical objects to finitely state certain bundles of facts about how physical magnitude properties and relations apply to non-mathematical objects.

14 Physical Magnitude Statements and Sparsity

Now let's return to the Classic Quinean Indispensability challenge and one of the most serious and influential motivations for it: suspicion that the Platonist can attractively logically regiment statements about physical magnitudes (like length and charge) while the Nominalist cannot.

In this chapter I'll review Putnam's famous cardinality argument that the Nominalist can't adequately paraphrase physical magnitude claims, and a follow up to it, which I will call the sparse magnitudes problem. Then I'll argue that we can evade these arguments and probably (technically) answer the relevant Quinean indispensability challenge by combining the modal if-thenist paraphrase strategy of Chapter 12 with two cheap tricks.

Interestingly, however, this paraphrase provides limited help to the Nominalist overall. For the answer to classic Quinean indispensability worries I propose is intuitively unsatisfying in certain ways, and considering it highlights the importance of associated Grounding and Reference indispensability arguments, which remain unanswered. However, I'll suggest that merely clarifying the nominalist's real problem with physical magnitudes in this way has some philosophical value.

14.1 Putnam's Counting Argument

Hilary Putnam (1971) makes an influential counting argument that (a certain kind of) Nominalist cannot write logically regimented sentences which are "adequate for the purposes of science" because they cannot (appropriately) logically regiment certain statements about lengths.

In particular, Putnam targets a materialist Nominalist, who believes in broadly material objects like sticks, stones and electrons, but not any immaterial objects like numbers or spatial points. He notes that many scientific theories are ordinarily stated by appealing to a physical magnitude function, like a length or mass function which relates physical objects to numbers. For the Nominalist to capture the way we apply laws like Newton's law of gravity:

$$F = \frac{gM_aM_b}{d^2}$$

they must interpret length statements that have (something like) the following form:

L_i: "c is $q_0 \pm q_1$ times the length of d"[1]

where q_0 and q_1 are rational numbers.[2]

Now Putnam argues that his Nominalist cannot adequately formalize such statements because (roughly speaking) they can't accommodate the intuition that it's epistemically possible for objects to stand in arbitrary length ratios to one another even while there only, say, less than 3000 total material objects (hence less than 3000 total objects, from their point of view).

Intuitively, a pair of sticks could (in principle) stand in infinitely many different length ratios to one another, while existing in a world with $n \geq 2$ or fewer material objects. Indeed, for each pair of length statements $L_i \neq L_j$ like "The ratio between c and d is 3.2 ± 1" and "The ratio between c and d in 4.1 ± 3.37," it's metaphysically and (initially) epistemically possible that one is true. Furthermore, we shouldn't rule out the possibility that there are at most n material objects, but $\neg(L_i \leftrightarrow L_j)$ (for any distinct L_i and L_j). So, from their point of view, no statement of the following form (for distinct sentences L_i and L_j) is a necessary truth:[3]

Finite Objects Conditional for L_i, L_j: "If the number of individuals is at most n then $L_i \leftrightarrow L_j$."

However, we can show by a counting argument that for some i, j the associated Finite Objects Conditional must be a necessary truth (assuming, as is commonplace, that all statements are formalized using only finitely many[4] relations $N_1, \ldots, N_m$). For there are only finitely many distinct scenarios which the application of $N_1, \ldots, N_m$ to at most n objects can distinguish. But there are infinitely many distinct length ratio sentences (i.e., sentences of the form L_i for some i). Hence, by the pigeonhole principle, some pair of distinct length sentences L_i and L_j must take on the same truth-value in all such scenarios. Thus, the nominalist's rendering of some Finite Objects Conditional with distinct L_i and L_j must be a necessary truth, violating the intuitive epistemic possibility requirement.

In contrast, Putnam argues that a Platonist who believes in spatial points (or paths) can attractively regiment physical magnitude statements in a way that allows for the epistemic possibility that the sticks could stand in any of infinitely possible length ratios. His idea is essentially[5] the following. One can define the length ratio between path p_1 and p_2 by saying that the ratio of the length of p_1 to p_2 is r iff for every function f which respects[6] the nominalistic relations:

---

[1] I mean this to abbreviate a corresponding claim where c and d are replaced by nominalistically acceptable definite descriptions or names of spatial paths (or other physical objects) with lengths.

[2] Note that by using only rational numbers, but including a margin of error, Putnam is able to approximate the claim that x has length r times the unit meter stick in Paris (though, of course, now the meter is defined differently) for any real number r to any degree of accuracy (while speaking a finite human-learnable language).

[3] More specifically, Putnam points out that they shouldn't regard any sentence of the form below as a theorem.

[4] C.f Davidson (1967).

[5] Putnam actually defines a "distance in meters" function which applies to pairs of points. My deviation from Putnam's approach is inspired by Sider (n.d.).

[6] The function $l(x)$ respects $\leq_L$, $\oplus_L$ just if $a \leq_L b \leftrightarrow l(a) \leq l(b)$ and $\oplus_L(a,b,c) \leftrightarrow l(a) + l(b) = l(c)$.

$p_1 \leq_L p_2$ "path p_2 is as long or longer than path p_1"
$\oplus_L (p_1, p_2, p_3)$ "the combined lengths of path p_1 and p_2 together are equal to the length of path p_3"

and satisfies certain other obviously necessary conditions for being a correct length function, we have $f(p_1)/f(p_2) = r$. Measurement theoretic uniqueness theorems ensure that if space satisfies the condition that length is **richly instantiated** (see Appendix D) then there is a unique function f satisfying these conditions (and thus length-ratios are all well-defined). The assumption that length is richly instantiated is, roughly, a way of asserting that any path can be subdivided into n equal length subpaths (note, we will later use the fact that such a condition can be nationalistically stated).[7]

14.1.1   Responding to the Counting Argument

As work like Field (1980) has emphasized, Putnam's original argument about lengths doesn't stymie a Nominalist who accepts spatial points (as the set-theoretically motivated Nominalist we're considering (c.f., Chapter 10) is perfectly free to do).

Indeed, we can directly use the paraphrase strategy from Chapter 12 to nominalize Putnam's own formalization of physical magnitude statements. For the latter only makes use of the reals, natural numbers, functions from paths to reals and functions from pairs of paths to reals (all structures which clearly definably supervene on facts about the application of Nominalist vocabulary) and the nominalistic relations "is a path," $p_1 \leq_L p_2$ (path p_2 is as long or longer than path p_1) and $\oplus_L(p_1, p_2, p_3)$ (the combined lengths of path p_1 and p_2 together are equal to the length of path p_3).[8]

---

[7] One might wonder what becomes of the counting argument if we take this line. It's striking, and prima facie puzzling, fact that Putnam's counting argument for the "false theorem" that the Nominalist must accept doesn't make use of any premises that are specific to the Nominalist. So, it looks like the Platonist is equally committed to accepting the fact that some version of the Finite Objects Conditional must be a necessary truth. How does the view Putnam ultimately proposes avoid falling to the same criticism that leads Putnam to reject the type of Nominalism he is rejecting?

I take the solution to this mystery to be something like the following. Philosophers who accept spatial points don't have to regard this conclusion (saying "If the number of individuals is at most n then $L_i \leftrightarrow L_j$" for distinct L_i and L_j) as a false theorem. For, they will say that all the intuitively epistemically possible scenarios considered involve infinitely many spatial points. Thus, accepting the truth of $L_i \leftrightarrow L_j$ in possible worlds with only n or fewer total objects is no problem.

To put the same point differently, what's *actually* a "false theorem" is this: If there are $< n$ many physical objects, then a pair of objects stands in ratio L_i if and only if they stand in ratio L_j (nothing can have a mass ratio in such and such range).

Putnam's materialist has a problem because they can deduce this claim from the claim Putnam proves – that if there are $<_n$ many material objects (where this doesn't include spatial points) then all things stand in length ratio L_i iff they stand in length ratio L_j.

So, the answer to my question is that Putnam's Platonist avoids trouble because they reject this inference.

[8] Thus we can apply the results from Chapter 12 to show that, where ϕ is a length statement logically regimented in the way that Putnam proposes, $T(\phi)$ is defined and gets the right truth-value in all metaphysically possible worlds if ϕ does:

Thus, Putnam's argument about nominalistically formalizing length statements can be answered. One might hope to generalize this strategy by treating mass, charge, etc. statements the same way. However, an important problem has been raised for generalizing this response.

14.2   The Sparse Magnitudes Problem

Considering physical magnitudes other than length raises a potential problem for a Nominalist using the strategy from Section 14.1. Specifically, it seems some physical magnitudes can take on definite values which don't supervene on the nominalistic relations between objects with these physical magnitudes. As Eddon (2013) puts it:

It seems possible for there to be a world, w_1, in which a and b are the only massive objects, and a is twice as massive as b. It also seems possible for there to be a world, w_2, in which a and b are the only massive objects, and a is three times as massive as b. Worlds w_1 and w_2 are exactly alike with respect to their patterns of LESS and CONG relations. And thus they are exactly alike with respect to the constraints these relations place on numerical assignments of mass. But if they are exactly alike with respect to the constraints these relations place on numerical assignments of mass, then it cannot be the case that these worlds differ with respect to the masses of a and b. So it seems we cannot discriminate between the two possibilities we started out with.[9]

Field notes and discusses a version of this problem in Field (1984) and the last chapter of Field (1989).

We can turn Eddon's point more into a more general and explicit argument that no analysis of a certain kind could work, by formulating it as a slight modification of Putnam's counting argument. Consider mass ratio sentences of the form

M_i: "The masses of objects c and d stand in ratio $q_{i_0} \pm q_{i_1}$."

By the intuition Eddon expresses in the quote above, it seems that, for each number $n \geq 2$ and mass ratio $r \in \mathbb{R}$, it is (metaphysically and, for large values of n, epistemically) possible that only n objects that have masses exist, and yet the objects c and d stand in mass ratio r. Given n, for any distinct mass ratio statements M_i and M_j, it is possible for there to be at most n objects but $\neg(M_i \leftrightarrow M_j)$. So, by Putnam's argument, any attempt to regiment M_i using finitely many relations **between the objects that have masses** will fail.

$$T(\phi) \overset{\mathrm{def}}{\leftrightarrow} \Box \underset{L}{\leq}, \underset{L}{\oplus}, \mathrm{path}\, (D \to \phi)$$

where D formalizes the claim that (fixing $\leq_L$, $\oplus_L$, path) there are objects satisfying a categorical description of the reals as well as objects corresponding to all possible functions from pairs of paths to reals above. And ϕ is the Platonist paraphrase for the claim that all functions f satisfying Putnam's conditions also make L_i true, i.e., $f(x)/f(y)$ is in $q_{i_0} \pm q_{i_1}$.

[9] This is stated as an objection to a nominalization attempt involving only two physical magnitude relations, but the same counterexample works if we add the third $\oplus$ relation considered in Section 14.1.

All this raises a worry for the Nominalist. For it suggests that we cannot be sure of producing an adequate nominalistic paraphrase if we treat mass statements the same way we treated length statements Section 14.1. Unlike in the case of length, mass facts intuitively don't supervene on how the relations

$p_1 \leq_M p_2$ iff path p_2 is as massive or more massive than path p_1
$\oplus_M (p_1, p_2, p_3)$ iff the combined masses of path p_1 and p_2 together are equal to the mass of path p_3

apply to physical objects that have masses.

14.3   Cheap Tricks

In this section, I'll argue that we can plausibly solve this problem (sufficiently to answer classic Quinean indispensability worries), by supplementing the paraphrase strategy of Chapter 12 with a pair of cheap tricks.

14.3.1   Relation to Length Trick

First note that if we (temporarily) assume that *length* is richly instantiated, we can solve Eddon's problem by using distance ratios to nominalistically pin down other physical magnitudes.

Specifically, a Nominalist who wants to paraphrase physical statements' masses, or any other property (given by real numbers), can do so by invoking a four-place relation M between pairs of objects with masses and pairs of paths, as follows:[10]

$$M(p_1, p_2, m_1, m_2)$$

which holds iff the mass of m_1 is as many times (or more) the mass of m_2 as the length of the path p_1 is to the length of the path p_2.

Even though this four-place relation may not be very physically (or metaphysically) natural, it reflects a genuine nominalistically acceptable fact about the world and suffices for our purposes. By the measurement theory results mentioned in Section 14.1, we can uniquely pin down the length function (up to a choice of unit) using Nominalist relations $p_1 \leq_L p_2$ and $\oplus_L(p_1, p_2, p_3)$, at all worlds where length is richly instantiated and then pin down mass ratio facts using M, which compares them to length ratios.[11] This, in turn, is enough to allow us to apply the paraphrase strategy of Chapter 12.

[10] If you prefer to take length to relate pairs of spatial points rather than paths, as Field's strategy for nominalistically stating rich instantiation conditions requires, we can replace each path with a pair of spatial points.

[11] Specifically, we demand that any mass function $\mathcal{M}$ satisfy the constraint that if L is a length function respecting $\leq_L$, $\oplus_L$ then $M(m_1, m_2, p_1, p_2)$ holds iff $\mathcal{M}(m_1)/\mathcal{M}(m_2) \geq \mathcal{L}(p_1)/\mathcal{L}(p_2)$. Note that any attempt to assign the wrong mass ratio r' to a pair of objects m_1, m_2 with mass ratio r can be ruled out by considering paths p_1, p_2 whose length ratio falls between that of r and r' and noting that M fails the above condition for a pair of paths such that $\mathcal{L}(p_1)/\mathcal{L}(p_2)$ falls between r and r'. The existence of such a pair of paths is guaranteed by the assumption that length is richly instantiated.

Importantly, even if length isn't *necessarily* richly instantiated, the paraphrases proposed by our paraphrase strategy still give the correct truth-values in those worlds where length is richly instantiated.[12]

14.3.2   Holism Trick

What about the assumption that length is *metaphysically necessarily* richly instantiated? This assumption seems unmotivated but, happily, we can eliminate it via another cheap trick if we are substantivalists about space and (as currently appears to be the case) our best scientific theory implies that length is *actually*[13] richly instantiated (as a matter of physical law).

To see how, consider some such Platonistic theory P that implies that space is richly instantiated. As noted Section 14.3, we can produce a partially accurate nominalization (call it $T^*(P)$), which gets the correct truth-value at worlds where length is richly instantiated but may get the wrong truth-value at other possible worlds. And we can formalize the claim that length is richly instantiated nominalistically, as per Appendix D, with a sentence R.

Thus, we can write a nominalistic sentence which get correct truth values at *all* possible worlds by simply conjoining our partial paraphrase of the theory $T^*(P)$ with the claim that length is richly instantiated R:

$$\text{Paraphrase: } T^*(P) \wedge R$$

At worlds where length is richly instantiated, $T^*(P)$ has the correct truth-value, and R is true, so their conjunction is true; and at worlds where length isn't richly instantiated, the theory we're paraphrasing is false, and so is R. Thus, our paraphrase has the intended truth-value at all metaphysically possible worlds.

This strategy can be easily extended to handle physical magnitudes taking values in $\mathbb{R}^n$ or $\mathbb{C}^n$. In fact, this second "cheap trick" isn't needed. For with a little reasoning we can show that $T^*(P)$ itself is false at all possible worlds where length isn't richly instantiated, hence that (the Platonist must allow that) $T^*(P)$ is true at exactly the same possible worlds as $T(P)$.[14] This provides a uniform paraphrase strategy that correctly

---

[12] If desired, we can use the same trick to ensure that determinate values for all physical magnitudes can be pinned down in all possible worlds where *at least one physical magnitude* is richly instantiated (whether or not it's length). Rather than requiring that physical magnitude functions honor a single four-place relation claim M (like the one relating length and mass statement above) require that it honor a corresponding four-place relation for each pair of distinct physical magnitudes M_i and M_j as well as respecting the particular $p_1 \leq_{M_i} p_2$ and $\oplus_{M_i}(p_1, p_2, p_3)$ relations for each of these magnitudes.

[13] For example, we can make the same argument below if our best theory implies that space is quantized, i.e., all distances between spatial points are multiples of some minimum distance. The same argument given by Sider (n.d.) to argue for measurement theoretic uniqueness results and the same tricks useable to nominalistically state the claim that space is Archimedean turn out to work in this case.

[14] Thanks to an anonymous reviewer for this point. When P implies that length is richly instantiated, we can show that show that $T^*(P)$ is already false at all worlds where length is not richly instantiated. For if length is not richly instantiated, then it's logically necessary given the facts about how all relevant nominalistic relations $N_1, \ldots, N_m$ (which include the length ratios relations) apply that length is not richly

paraphrases a wider range of claims, e.g., this strategy correctly nominalizes many isolated scientific claims as well as complete theories (as per the aim of answering Quine by uniformly nominalizing a language adequate for an overall scientific practice discussed in Section 11.3.2).

Thus, the Nominalist plausibly *can* address the problems about physical magnitude statements raised by Putnam and Eddon – sufficiently well to answer the classic Quinean indispensability argument.

14.4   Remaining Indispensability Worries

However, readers will likely find the solution of the previous section quite unsatisfying! I'll suggest that attending to our discomfort highlights further grounding and reference indispensability worries of the type foreshadowed in Section 11.4.

14.4.1   Grounding Indispensability

First, it feels worrying that the kind of four-place relation M (between paths with lengths and pairs of objects with masses or some other target physical magnitude) invoked in Section 14.3.1 is so extrinsic, arbitrary and not physically natural. It's (prima facie) very much not the kind of relation we'd like to make part of our fundamental ideology when doing metaphysics.

Thus, a Nominalist who addresses the classic Quinean indispensability argument via the tricks I've proposed in this chapter, arguably faces a grounding problem. If the four-place relation we used to paraphrase mass facts isn't metaphysically fundamental, what kind of metaphysically fundamental objects and relations *do* ground the fact that m_1 is π times more massive than m_2? Platonists can say that mass facts are grounded in a relation between physical objects and mathematical ones, e.g., the three-place relation which holds between pairs of objects and their real-valued mass-ratio. But what can the Nominalist say?

---

instantiated. So we have $\neg R$ and $\Box_{N_1,\ldots,N_m}(D \rightarrow \neg R)$. By assumption, P implies that length is richly instantiated. So $(P \wedge D) \rightarrow R$ is a logical truth. Thus, we can infer $\Box_{N_1,\ldots,N_m}(D \rightarrow \neg P)$ (i.e., the claim that $T^*(\neg P)$ is true): Lemma B.2 (Box Introduction). To get the claim that $T^*(P)$ is false, note that it's logically possible to satisfy the platonist's assumptions D about how non-mathematical reality and mathematical objects relate, at every metaphysically possible world. So, at the worlds in question, we have $\Diamond_{N_1,\ldots,N_m}D$. By importing and applying the fact that $\Box_{N_1,\ldots,N_m}(D \rightarrow \neg P)$ inside this $\Diamond_{N_1,\ldots,N_m}$ context, we can deduce that $\Diamond_{N_1,\ldots,N_m}(D \rightarrow \neg P)$. Suppose, for contradiction, that $T^*(P)$ was true at these worlds. Then we'd have $\Box_{N_1,\ldots,N_m}(D \rightarrow P)$, which is equivalent to $\sim \Diamond_{N_1,\ldots,N_m}(D \wedge \neg P)$. Thus $T^*(P)$ is false at all worlds where length is richly instantiated.

Employing this strategy for more fully responding to Quinean challenges involving physical magnitude statements might require further philosophical and technical work generalizing the basic paraphrase strategy described in Chapter 12 to handle cases where our would-be definable supervenience description D doesn't pick out a unique Platonist structure at some worlds (perhaps generating truth-value gaps at these worlds). But I won't say more about it here.

14.4.2   Reference Indispensability

Second, one might feel that nominalistically formalizing our best scientific theory via the cheap tricks described in this chapter leaves our apparent ability to make certain other claims a mystery. For example (rejecting holism) one might think that we can – somehow – express mass statements on their own, rather than conjoined with a larger physical theory. Or one might think we can meaningfully state mass claims conjoined with alternative larger physical theories which *don't* imply that length is richly instantiated. And the cheap tricks I've suggested in this chapter don't suffice – or don't clearly suffice – to let one nominalistically formalize such statements.

Thus, one might argue that mathematical Nominalists can not account for our *ability to state* such partial and/or alternative scientific theories. In contrast, Platonists can make sense of them by taking our language to include something like a three-place relation "*x* stands in mass ratio *r* to *y*" between physical objects and mathematical ones. Thus, we might be forced to accept the existence of mathematical objects in order to make sense of our ability to literally state certain, seemingly meaningful, scientific theories *other than* our current best theory.

I have called this a reference explaining worry because it concerns accounting for our ability to (finitely learn a language which lets us) "refer to" certain classes of possible worlds by stating sentences that are true at exactly these worlds.

14.4.3   Morals

In this chapter I've argued that a (set theoretically motivated) nominalist can plausibly answer classic Quinean indispensability worries about adequately formalizing physical magnitude statements in our best scientific theories of the world, by combining the paraphrase strategy of Chapter 5 with some cheap tricks.

However, Putnam and Eddon's points still highlight a prima facie problems for the Nominalist because a significant reference and grounding indispensability worry about physical magnitude statements remains. Although not needed to state our best scientific theories, mathematical objects may be indispensable to accommodating certain philosophical intuitions about reference and grounding. If there are no numbers, how are humans able to finitely learn languages which draw certain distinctions between metaphysically possible worlds quite different from our own (not needed to state our best theory), and what could ground the truth of fundamental physical magnitude facts in the worlds?

If one accepts this reformulation of indispensability worries, there are some important upshots for readers of different stripes. First, hardcore naturalists may be inclined stop taking indispensability worries (based on concerns about physical magnitude statements) seriously. For we see that the nominalist's real problem doesn't concern stating or (in a sense) attractively explaining *scientific* facts involving mass and charge but rather accounting for certain a priori philosophical intuitions about metaphysical possibility, reference and grounding. Philosophical explanation is the sticking point, not scientific explanation.

Second, if we slot the grounding and reference indispensability arguments into the philosophical literature in place of the classic Quinean indispensability argument but take the latter seriously (accepting the existence of mathematical objects on these grounds) there are several interesting consequences.

For one thing, if one accepts the existence of mathematical objects because of the grounding challenge in this chapter, then one has an automatic answer to certain access worries. I have in mind the suggestion (Jenkins 2008) that if there hadn't been mathematical objects everything about the physical world would have been the same. For, if masses are grounded in (and thus, plausibly, something like partly constituted by) a certain relation holding between physical objects and numbers, then the following (opposite) counterfactual intuition seems plausible: if numbers suddenly stopped existing, then objects wouldn't have had masses, just as if hair suddenly stopped existing then people would stop having beards.

For another thing, consider arguments that we're only justified in believing mathematical objects exist because of the role they play in our best scientific theories (as per the Quinean indispensability argument) (Colyvan 2001), so we should think mathematical objects are contingent. The reference and grounding indispensability arguments raised in in this section present a twist on the classic Quinean indispensability argument which (if compelling) does justify the necessary existence of mathematical objects. For, in order to resolve the grounding and reference problems raised in this chapter, mathematical objects would seem to need to exist necessarily.

15 Weak Quantifier Variance and Mathematical Objects

15.1 A Different Approach to the Unity of Mathematics

So much for Nominalism. Now let's turn to the philosophy of mathematics I ultimately want to advocate.

In Chapter 10, we considered a "unity of mathematics" argument that philosophers who accept the Potentialist set theory advocated in this book should also be Nominalists about other mathematical objects, cashing out all of pure mathematics as an investigation of pure logical possibility facts (i.e., claims stateable without any quantifiers occurring outside of unsubscripted $\Box$ and $\Diamond$ claims). I suggested (very briefly) that a neo-Carnapian approach which combined realism about mathematical objects outside of set theory with Potentialist set theory could also honor these mathematical uniformity intuitions.

In this chapter, I will develop such a proposal, which I call the Weak Quantifier Variance Explanation of Mathematicians' Freedom (QVEMF). It appeals to a Weak Quantifier Variance thesis in philosophy of language, which takes inspiration from some familiar and popular neo-Carnapian ideas, but (as we will see) does not require endorsing controversial Carnapian rejections of metaphysics.

As noted in Chapter 10, adopting such a view lets us honor Benacerraf's idea that we should treat apparently grammatically and inferentially similar talk of numbers, cities and electrons similarly, and avoid classic Quinean and Reference explaining challenges, since we acknowledge that mathematical objects literally exist. However, in Section 15.3 I will note that a kind of Grounding Indispensability challenge arises and discuss some ways of avoiding or answering this challenge.

In Chapter 16 I will consider some more general concerns about developing and defending Weak Quantifier Variance.

15.2 Weak Quantifier Variance and Mathematicians' Freedom

15.2.1 Motivations

To motivate and introduce the Weak Quantifier Variance thesis which supports the approach to mathematics I want advocate, consider our knowledge of holes and shadows.

In ordinary contexts we appear to quantify over objects like holes in a road or in a piece of Swiss cheese. For example, we may say that there are three potholes in the road between one town and another, or that one piece of cheese has more holes in it than another. And if one accepts the existence of these holes, it is appealing to think of them as distinct from things like the air that occupies them or surrounding portions of the "hole host" (e.g., the cheese or the pavement).[1]

Is there an access problem about our knowledge of holes? One might try to get such an access worry going, by arguing as follows. Our ability to visually determine how many holes there are in a road, depends on our (implicit or explicit) accuracy concerning how hole facts supervene on facts about the distribution of solid matter in space.[2] For example, we must be disposed to make correct judgments about how steeply indented a road must be to count as containing a hole. But what can explain the match between our beliefs on this topic and the corresponding objective reality about when there is a hole in the road? It doesn't seem like sensory experience or scientific practice strongly motivates thinking that any particular place to draw the line is intrinsically physically/metaphysically special (even allowing for some vagueness). Thus, people's apparent ability to draw the line correctly (re: how steeply indented a substance has to be to contain a hole) could seem to create an access problem.[3] Consider the match between *facts* of the form below and human *beliefs* about these facts:

When a road is missing a cylinder of material of depth 3 cm and width 15 cm, there is a hole in that road.
    When a road is missing a cylinder of material of depth 0.01 cm and width 0.1 cm, there is *not* a hole in that road.

We process visual information in way that draws the line somewhere (maybe with some vagueness), but what explains the match between where we do draw the line and the correct place to draw the line?

However, it's appealing to say that there isn't really any such access problem for holes, because one can give the following metasemantic explanation for human accuracy about minimum hole indentation facts and the like. If we had been inclined to say something (logically coherent but) different about when an indented object counts as "containing a hole" (e.g., that substances surfaces had to contain an indentation of greater/lesser steepness in order to contain a hole) then the meaning of the words "hole" and "there is" would have been different, so that our utterances would have still expressed truths. That is, we would have been speaking as slightly different language in which a slightly different collection of sentences of the form "Whenever a solid road is indented according to a geometrical formula ϕ, there is a hole in it"

---

[1] See Lewis (1990).   [2] Though see Berry (2019a) for a puzzle about this notion of solidity.
[3] Note that the issue with this "access problem for holes" is not supposed to be about vagueness, but about our ability to be accurate (or even close to accurate) about how hole facts supervene on indentation facts.

express true propositions.[4] Accordingly, there's no mystery or spooky Leibnizian predetermined harmony in our possession of true beliefs about things like about how steeply indented holes must be.[5]

Note that this explanation seems to involve (a form of) quantifier meaning change, in that it requires that our adopting different hole attribution practices would have caused a shift in the meaning of some of our logical vocabulary like the existential quantifier (not just a shift in the meaning of the world "hole"). For example, note that changing between more and less generous standards for hole existence could require the truth-value of the Fregean sentence which says "There are n things" using only first-order logical expressions and equality.[6]

15.2.2   Introducing Weak Quantifier Variance

Thinking about cases like those in Section 15.2.1 motivates the following Weak Quantifier Variance Thesis:

(Weak) Quantifier Variance Thesis:
- There are a range of different meanings "there is" could have taken on, which all obey the syntactic rules for existential quantification.[7]
- These senses need not all be mere quantifier restrictions of some fundamental maximally natural quantifier sense (if there is one).[8]

I call this claim the *Weak* Quantifier Variance thesis because it doesn't include a further "parity" claim (that none of these variant quantifier senses is somehow metaphysically

---

[4] Arguably our current language allows for contextual variation in how strict the standards for hole existence are and hence (for the reasons to be discussed below) corresponding variation in the meaning of "there is." So one might think of there being a shared core meaning to "there is" (perhaps associated with the introduction and elimination rules) which combines with contextual factors to determine truth conditions for sentences involving "there is" at each metaphysically possible worlds. For present purposes I'll simply talk about shifts in quantifier meaning, but I don't mean to prejudge this issue.

[5] Or at least there's no mystery if we bracket access worries about knowledge of logical coherence.

[6] For example, the sentence that says there are two things $(\exists x)(\exists y)[\neg x = y \wedge (\forall z)(z = x \vee z = y)]$.

   Also note that the quantifier meaning change involved explanation above does not suggest that when we start talking in terms of holes and shadows (or switch from stricter to laxer standards for hole existence) we bring these objects into being. The existence of holes and shadows is not caused by, or grounded in, the existence of language users who talk in terms of holes and shadows, and it will be true to say "there were holes before there were people, and before I started talking in terms of them." Instead we are merely changing our language so that some sentences, e.g., "there is something [namely, a hole] in the region of the cheese plate which is not made of matter" go from expressing a false proposition in our old language to expressing a different, true, proposition in our current language (see Einheuser (2006) for a vigorous development of this point).

[7] By this I mean that, for each such quantifier sense there is some possible language such that all applications of the standard syntactic introduction and elimination rules for the existential quantifier within that language are truth preserving. However, that does not mean that one can form a single language containing both quantifier senses and then apply the introduction and elimination rules to prove the equivalence of these senses. See Warren (2014), among others, on this point.

[8] That is, these variant quantifier senses need not be interpretable only as ranging over some subset of the objects which exist in the fundamental quantifier sense, in the way that we might say the "all" in a typical utterance of "all the beers are in the fridge" restricts a more generous quantifier sense to only range only over objects in the speakers' house.

special) which is generally included in definitions of Quantifier Variance.[9] So, for example, it would be compatible with Weak Quantifier Variance to say that there's a maximally natural quantifier sense corresponding to what objects exist fundamentally.

And, indeed, some friends of traditional metaphysics have found their own reasons for accepting the Weak Quantifier Variance thesis (and thereby putting themselves in a position to give the Quantifier Variance Explanation of Mathematicians' Freedom defended in this chapter). For example, Sider (2009) uses Weak Quantifier Variance to capture the intuition that ordinary speakers' non-philosophical utterances like "There's a hole in the road" can express uncontroversially true statements, even if it's an open question whether holes exist in the sense more relevant to the (traditional fundamental) metaphysics seminar. Sider says there's a unique, maximally natural, sense of the quantifier which ontologists aim to study/employ.[10] But he allows that there are also other (perhaps less metaphysically joint-carving) senses, which the quantifier can take on in ordinary contexts, on which utterances of "There is a hole in the road" clearly can express a true proposition.

Note that saying some kinds of objects (e.g., cities, numbers) might not exist in the sense relevant to the Sider's fundamental ontology room doesn't amount to saying that these objects "don't really exist." It is entirely compatible with truthful assertion that these objects literally exist in the course of daily life (and while studying ethics or non-fundamental metaphysics about money and gender, or writing philosophy of mathematics books like this one) – much as acknowledging that rabbits don't exist on the (relatively) more natural and joint-carving quantifier sense employed by fundamental physics is compatible with saying rabbits literally exist in most ordinary contexts, including biology seminars. When outside the fundamental physics/ontology room, our position on such objects seems much more naturally expressed by saying that rabbits/holes/cities/numbers *might not be fundamental* than that they *don't really exist.*[11]

---

[9] See, for example, Eklund (2009), Hirsch (2010) and Chalmers' characterization of Quantifier Variance as (roughly) the idea that, "there are many candidate meanings for the existential quantifier (or for quantifiers that behave like the existential quantifier in different communities), with none of them being objectively preferred to the other" (Chalmers 2009).

[10] See the argument that (even from Sider's point of view) we don't *actually speak* a language with Sider's maximally joint-carving quantifier sense in most philosophical contexts (including discussions of metaphysics and ontology).

[11] Also note that (as discussed in Berry (2015)) accepting the Weak Quantifier Variance Thesis does not require one to accept that normal English employs verbally different expressions corresponding to at least two different quantifier senses (a metaphysically natural and demanding one and a laxer one) at the same time, so that it might be true to say things bad-sounding things like "composite objects exist but they do not really exist" in certain contexts. With regard to any particular context, we can fully agree with David Lewis that, "The several idioms of what we call 'existential' quantification are entirely synonymous and interchangeable. It does not matter whether you say 'some things are donkeys' or 'there are donkeys' or 'donkeys exist.' whether true or whether false all three statements stand or fall together" (Lewis 1990).

15.2.3 Explaining Mathematicians' Freedom

Now let's turn to the special case of mathematics. Contemporary mathematical practice seems to allow mathematicians significant freedom to introduce new kinds of mathematical objects, such as complex numbers, sets and the objects and arrows of category theory. For example, Cole (2013) writes:

Reflecting on my experiences as a research mathematician, three things stand out. First, the frequency and intellectual ease with which I endorsed existential pure mathematical statements and referred to mathematical entities. Second, the freedom I felt I had to introduce a new mathematical theory whose variables ranged over any mathematical entities I wished, provided it served a legitimate mathematical purpose. And third, the authority I felt I had to engage in both types of activities. Most mathematicians will recognize these features of their everyday mathematical lives.

Philosophers of mathematics face a challenge about how to account for this, and they have developed a number of styles of response.[12]

If we accept the Weak Quantifier Variance Thesis, we can explain mathematicians' freedom to introduce new kinds of apparently coherent objects along the following lines:

Quantifier Variance Explanation of Mathematicians' Freedom: When mathematicians (or scientists or sociologists) introduce axioms characterizing new types of objects, this choice can not only give meaning to newly coined predicate symbols and names but can change/expand the meaning of expressions like "there is," in such a way as to ensure the truth of the relevant hypotheses. Thus, for example, mathematicians' acceptance of existence assertions about complex numbers might change the meaning of our quantifiers so as to make the sentence, "There is a number which is the square root of –1" go from expressing a falsehood to expressing a truth. Similarly, sociologists' acceptance of ontologically inflationary conditionals like, "Whenever there are people who ... there is a country which ..." can change the meaning of their quantifiers so as to ensure that these conditionals will express truths.[13]

Hitherto, I take it, versions of QVEMF have largely been developed by philosophers who combine acceptance of the Weak Quantifier Variance thesis with some strong antimetaphysical claim (such as the parity claim referenced in Section 15.2.2) or project.[14] However, I'm suggesting that more metaphysically realist philosophers could also adopt QVEMF (backed by the Weak Quantifier Variance Thesis) and should consider doing so.

Adopting the Weak Quantifier Variance Thesis and accepting the existence of mathematical objects together with the QVEMF explanation for mathematicians' freedom lets us honor Benacerraf's goal of treating apparently grammatically and inferentially similar talk of numbers and cities similarly (acknowledging that both apparent kinds of objects exist). It allows us to say that a single notion of existence is relevant to claims like "Evelyn is prim" and "Eleven is prime" in any given context

---

[12] I will say something about how we might generalize this to the case of applied mathematical knowledge (like principles concerning sets with ur-elements) in the next chapter.

[13] See Berry (2015, 2019b).

[14] Here I have in mind Rayo (2015) and Thomasson (2015) as well as Hirsch (2010).

(though, of course, future choices may further change which notion of existence one's language employs). Proponents of this view need not say that mathematicians' statements are literally false,[15] or say that mathematical statements have a different logical form from claims which ordinary speakers treat similarly (e.g., apparent existence claims about holes and countries), in cases where the specific reasons (like the Burali-Forti worries in Chapter 2) for not doing so.

Admittedly many questions can be raised about Quantifier Variance and the Quantifier Variance explanation of mathematicians' freedom, which I can't discuss at any length here. For example, what would happen if mathematicians simultaneously adopted a pair of internally consistent, but incompatible, conceptions of pure mathematical structures? What would happen if mathematicians adopted a conception of some mathematical structure which imposed undue constraints on the total size of the universe (e.g., a logically coherent collection of axioms describing a purported mathematical structure which imply that the total universe contains at most 100 things)?

In a nutshell, I think we can answer the first challenge by saying that mathematicians' actual (and claimed) freedom only allows a given mathematical community/context to employ any logically coherent *total collection* of conceptions of pure mathematical structures.[16] So, a proponent of the Quantifier Variance explanation of mathematicians' freedom can say that if mathematicians simultaneously employ a pair of incompatible conceptions of mathematical structures (in some context), (a) this would be an accident and (b) at most one of these conceptions of mathematical structures would express a truth.

We can answer the second challenge by noting that axioms characterizing pure mathematical objects always employ quantifiers that are implicitly restricted to some collection of pure mathematical structures (see the discussion of implicit quantifier restriction in pure mathematics in Berry (2018b)), so these conceptions cannot impose any restrictions on the total size of the universe.

However, a fuller answer to these challenges would fit these claims into a general metasemantic story which also yields attractive verdicts about our practice of talking in terms of objects like holes, cities, contracts, etc. I'll sketch such a story in the next chapter.

15.3   Grounding Indispensability Worries for QVEMF

Accepting that mathematical objects (outside set theory) literally exist lets my Quantifier Variantist dodge the classic Quinean and (finitary) reference Indispensability worries for Nominalists. However, the neo-Carnapian realism about mathematical objects I advocate is deeply similar to the forms of Nominalism

---

[15] Recall Lewis (1991) saying, "I am moved to laughter at the thought of how presumptuous it would be to reject mathematics for philosophical reasons. How would you like the job of telling the mathematicians that they must change their ways, and abjure countless errors, now that philosophy has discovered that there are no classes?"

[16] Thanks to Tom Donaldson for helpful discussion on this point.

discussed in Chapter 10 (in various ways noted in that chapter). And something can feel troubling about the idea that mere language change can dissolve such a difficult problem.

The Grounding Indispensability argument against the QVEMF below develops this intuition. Note that (as per the Siderian picture in Section 11.4.2) this argument takes some notion of grounding (not necessarily the same one) to apply to: facts, objects and relations. Thus, we can talk about both whether facts involving mathematical objects are grounding fundamental and whether relations like "x stands in mass ratio r to y" are grounding fundamental.

Grounding Worry for Quantifier Variantists: All facts can be grounded in terms of facts involving only fundamental objects.[17] And (one might think!) accepting the Quantifier Variance explanation of mathematicians' freedom requires saying that all logically coherent character-izations of mathematical structures are "on par." Thus, proponents of the QVEMF must either say that all possible logically coherent mathematical structures are metaphysically fundamental (contra the core intuitions used to motivate Potentialist set theory in Chapter 2 or that no mathematical objects are metaphysically fundamental (e.g., all mathematical objects' existence is grounded in modal facts about logical possibility in some way that implies). So, it should be possible to ground all facts involving mathematical objects in facts that don't involve math-ematical objects. But, what can ground facts about physical magnitudes if not a relation to numbers? Furthermore, the requirements for nominalistically grounding applied mathematical facts are very similar to those for nominalistically paraphrasing applied mathematics facts. So, Quantifier Variantist realists about mathematical objects face a grounding indispensability problem which is just as bad as the reference indispensability problem.

Thus, one might conclude that the arguments about physical magnitude statements pose a serious problem for the Quantifier Variance realist as well as for the Nominalist.

I will argue that there are a number of attractive strategies for responding to this worry. First, of course, you might argue that the nominalistic Reference and Grounding challenge are both solvable. For example, various philosophers have advocated accepting platonic physical mass, charge and other abstracta. If such platonic physical magnitude abstracta existed, they could be used to answer both finitary Reference and Grounding worries about physical magnitude facts.[18]

Second, you could argue that the Grounding challenge is answerable while the Reference challenge is not. Recall that we had independent reason for thinking formal constraints on grounding are quite different from those specified for adequate para-phrase in Section 11.3.2. The finiteness and learnability constraints on nominalistic paraphrase don't apply when providing grounding. So certain arguments that we can't "adequately" nominalistically regiment physical magnitude statements for the pur-poses of finite reference explaining challenge don't work when applied to grounding.

For positive examples of answers to the Grounding challenge which aren't answers to the Reference challenge, see Hellman's story about how physical magnitude facts could be grounded in facts involving infinitely many different length/mass/whatever atomic properties. One might also suggest that the "language of metaphysical funda-mentalia" is sufficiently different from the languages humans speak, that fundamental facts about the extent to which something is F need not be grounded in facts about

---

[17] c.f. Sider (2011)'s purity thesis.   [18] One could deploy the strategy.

whether or not some binary property or relation holds between objects at all. Maybe what's metaphysically fundamental is analog, where language is binary, so to speak.

Third, you could reject the demand for grounding all together. And fourth, you could reject the parity reasoning in Section 15.2.2 above, the idea that someone who gives a quantifier variance explanation of mathematical objects is committed to saying that no mathematical objects are among the metaphysical fundamentalia (none "exist" on Sider's maximally natural quantifier sense). In the rest of the chapter, I will argue that the latter two styles of answers are more appealing than they might at first seem.

15.3.1   Rejecting Grounding

As Quantifier Variance has traditionally been developed as part of a larger neo-Carnapian program which rejects traditional metaphysical questions as meaningless, I suspect that rejecting demands for grounding all together would be most popular response to the Grounding challenge among my fellow neo-Carnapians.

Admittedly, this rejection may seem to come with a cost. For the notion of grounding provides one way of fleshing out an enduringly appealing idea: that an apparently complex universe and variegated language can be explained in terms of a few simple notions. Advocates of the Sideran framework reviewed in Section 11.4.2 will say there's a single small collection of maximally fundamental concepts and kinds of objects, such that facts about them ground everything.

However, neo-Carnapians can and have honored the same idea in a different way, by saying that (in some sense) everything can be reconceptualized in terms of a conceptually parsimonious basis language, but there are a range of different equally good basis languages (perhaps making different choices of mathematical ontology) at issue. We appeal to something like Augustin Rayo's symmetric "nothing but" relation (Rayo 2015) or talk of conceptual re-carving. And we then say that reality is "simple" in the sense that all facts expressible in our language (and maybe some specified range of other languages) bear this nothing-but relation to facts in some simple "basis language."

If we adopt this strategy (i.e., cash out metaphysical parsimony intuitions in terms of something like Rayo's symmetric "just is" relation rather than a grounding relation), we won't say there's a unique correct choice of basis language (the point of the metaphor of basis vectors is that there are a number of different choices which are equally good for representing a given vector space). Rather we can say that a range of choices of basis language are equally capable of bringing out the unity and elegance underlying the diversity and variety we see in the world.

On this strategy, the neo-Carnapian could grant that Nominalists' problems answering the Reference Indispensability Challenge discussed in Chapter 14 reveal that we need to think about physical magnitudes in terms of a relationship to *some* abstract objects (be they numbers on their own, numbers identified with certain sets, or the abstract mass objects, when choosing a parsimonious basis language adequate for stating a simple Theory of Everything). But they could say that all these ways of thinking in terms of different abstract objects are equally good choices of a simple basis language for drawing all the meaningful distinctions we want to draw and showing how a simple shared reality lurks under the apparently complexity suggested by natural languages (by paraphrasing natural language statements into some simpler language).

In this way, the neo-Carnapian can claim to achieve whatever the traditional Platonist thinks they've achieved (in terms of the Siderean unifying ambition) by *grounding* everything in a few things by showing that everything *stands in a just is/ conceptual re-carving* etc. relation to a simple basis language.

They will take any traditional Platonist story about what the supposedly metaphysically fundamental objects and concepts are (backed up with the kind of systematic paraphrases of sentences in an apparently richer language with sentences in an apparently narrower one) and say: that's one acceptable basis language. Whatever range of dappled and variegated facts the Platonist thinks are grounded in these few simple facts can indeed be adequately conceptualized in terms of this more limited language/facts/ideology.

However (a neo-Carnapian of this stripe can say) a different basis language which replaced the pure mathematical structures which this paraphrase strategy appeals to with different ones that can do the same applied mathematical work (e.g., replacing appeal to a free-standing copy of the natural numbers or reals with a copy of the numbers inside the hierarchy of sets) would be equally illuminating and "metaphysically insightful" (to whatever extent the neo-Carnapian will grant the meaning of the expression). Any sufficiently expressive pure mathematical language can be combined with some small collection non-pure-mathematical vocabulary to form an adequate basis language.

Note that this idea that different choices of pure mathematical structures (with sufficient expressive power) are somehow philosophically/metaphysically on par[19] fits naturally with a point from the literature on Quinean empiricist answers to access worries about mathematical objects. This is the idea that Quinean indispensability considerations don't seem to justify belief in *any particular* mathematical structure, as different mathematical structures seem capable of doing the same work in regimenting our physical theories[20] and physicists don't seem to care much which ones are invoked.

15.3.2   Agnostic Platonism

Now I want to draw attention to a different, less familiar, style of approach to the Grounding challenge, which I'll call Agnostic Platonism. Friends of the Quantifier Variance explanation of mathematicians' freedom who *don't* share traditional Carnapian opposition to metaphysics can take a different line in responding to the Grounding worry (that would not be available to Nominalists).

Suppose we grant that the history of debate over Quine's indispensability argument suggests some mathematical objects are among the metaphysical fundamentalia. Proponents of the QVEMF can still resist the Grounding worry by rejecting the idea that QVEMF implies all coherent conceptions of mathematical objects must be metaphysically (as opposed to merely mathematically) on par, and thence the argument that no mathematical objects can be grounding fundamental.

---

[19] See Sections 16.3 and 16.4 for some caveats and a way of thinking this through more carefully.
[20] See, for example, Clarke-Doane (2012).

In slogan form, someone who accepts agnostic Platonism would say: maybe some mathematical structures are metaphysically special, but mathematicians don't care which ones those are, and they don't need to care in order to reliably form true mathematical beliefs and satisfy the epistemic aims of the project of pure mathematics!

Perhaps indispensability arguments suggest that *some* mathematical objects (capable of doing certain applied mathematical work) exist fundamentally. But, as noted in Section 15.3.1, these considerations don't seem to justify belief in any particular mathematical structure – as different mathematical structures seem capable of doing the same work in regimenting/grounding our physical theories.

Allowing (in response to indispensability worries) that some mathematical structures may be metaphysically fundamental might seem to raise access worries (over and above the access worries about access to facts about logical coherence which the QVEMF theorist already faces[21]). For although these worries can suggest the fundamentalia plausibly include *some* mathematical objects, we don't know (and perhaps can never know) *which*.

But the agnostic Platonist avoids this access problem by saying that getting mathematics right doesn't require guessing which mathematical structures are among the fundamentalia. Note that this idea (that reliably speaking the truth in mathematical ordinary language doesn't require knowing the right answer to corresponding metaphysical questions about fundamental ontology) mirrors what it is natural to say about our knowledge of holes, in the following sense. It may turn out to be the case that some particular hole-like notion (maybe the topological notion of holes) will be used in physics. But construction workers can draw the line where they want with regard to hole boundaries and reliably speak the truth without having to take any such stance regarding fundamental metaphysics.

One might object that a similar access worry arises with regard to metaphysicians' knowledge of which mathematical structures are grounding fundamental. However, we can answer this access worry by noting that there's no access to account for. Metaphysicians don't even *appear* to know very much about which mathematical structures are metaphysically fundamental!

At this point a reader sympathetic to conventional Actualist set-theoretic foundationalism might object: how can I endorse the arbitrariness-based criticism of Actualist set theory developed in Chapter 1, while advocating Agnostic Platonism about mathematical fundamentalia without hypocrisy? For isn't dividing up the pure mathematical objects into those with fundamental existence vs. those without just as arbitrary as saying that the hierarchy of sets just happens to stop at a certain point? And isn't being committed to arbitrariness in which mathematical objects are fundamental just as bad as being committed to arbitrariness in size of the total mathematical universe?

Even if this charge of hypocrisy were correct, I think the Quantifier Variantist view advocated in this chapter would still be an improvement on conventional set-theoretic foundationalism. For the arbitrary joint posited by the agnostic Platonist doesn't constrain acceptable mathematical practice, whereas that posited by classic set-theoretic

---

[21] See Berry (2018b) for an argument that these access worries about logical coherence are solvable.

foundationalism does. The agnostic Platonist need not admit any limits on which logically coherent pure mathematical structures mathematicians could choose to talk in terms of. For they don't think mathematicians can only introduce or study structures which are grounding fundamental. In contrast, the conventional set-theoretic Actualist foundationalist holds that any conception of a pure mathematical structure mathematicians could legitimately adopt must have an intended model within the Actualist hierarchy of sets (thus constraining the space of legitimate structures mathematicians could adopt).

However, I will now sketch a more aggressive defense against this charge of hypocrisy. If the other assumptions needed for my Weak Quantifier Variance realist to face access worries hold (i.e., we need to provide grounding, and mathematical objects are needed for that task) then it seems that everyone, not just the Agnostic Platonist, must admit that certain mathematical structures are special in that they play a role in grounding non-mathematical facts about the world (e.g., maybe length reflects a fundamental facet of reality and length facts require grounding in the real numbers).

So the Agnostic Platonist still has the advantage that it only requires us to posit that one special joint in the space of coherent conceptions of mathematical structures (specifying which particular mathematical structures play a role in grounding and/or constituting particular applied mathematical facts, e.g., facts about events and probability, or lengths) where the classic set-theoretic foundationalist is committed to positing two joints in reality (this joint, plus the joint determining where the hierarchy of sets happens to stop). That is, both philosophers will be committed to some kind of fact like "the pure mathematical objects which play roles in grounding physical facts are exactly the real numbers and three layers of sets over them." But the set-theoretic foundationalist will also be committed to a fact like "the hierarchy of sets just happens to stop at X point" (where that point is usually taken to be large enough to accommodate all sets used in our physical theories).

Moreover, it seems more plausible that facts about the fundamental laws of physics might provide an, as yet undiscovered, principled division between those mathematical objects which play a role in grounding applied mathematical facts and those which don't, than it does that some choice of a height for the hierarchy of sets will turn out to be principled.[22]

---

[22] Indeed, one might argue as follows. Applied mathematics hasn't seemed to motivate a unique choice of which mathematical structures exist, because (from a traditional Platonist point of view) the total collection of mathematical objects must do two jobs. It must make sense of applied mathematics *and* everything we could study in pure mathematics. Given this goal, it has seemed natural to consider both, e.g., both a free-standing real number structure and a copy of the real numbers within various larger structures, like the hierarchy of sets (containing objects for pure mathematical study), as candidates for mathematical reference within our best physical theories. And there's no uniquely natural choice of a collection of mathematical objects which does both jobs.

However, the agnostic Platonist does not expect fundamental mathematical objects to do both these jobs. (As noted above) they can take the truth of existence claims about pure mathematical objects to be grounded in something like facts about logical possibility. Thus, it seems more plausible that whatever aspects of our best physical theories make appeal to fundamental mathematical objects indispensable (if such there are) should suggest a unique choice of which mathematical structures to take to be grounding fundamental.

Thus, to summarize, I think the (admittedly prima facie strange) idea of saying that, although mathematicians can introduce any pure mathematical structure they like, some pure mathematical structures are metaphysically special and instantiated by objects which are grounding fundamental is more appealing than it first seems.

16 Weak Quantifier Variance, Knowledge by Stipulative Definition and Access Worries

In this chapter I will zoom out and further develop the Weak Quantifier Variance thesis. First, I'll use the paraphrase strategy from Chapter 12 to further explicate my statement of the Weak Quantifier Variance Thesis in Section 15.2 (eliminating informal appeal to languages that "talk in terms of more objects than our own"). In doing this, I hope to answer worries that neo-Carnapian theories can't be literally stated without making paradoxical claims like "there is something that I'm not now quantifying over".

Second, I'll discuss a common worry about the kind of neo-Carnapian response to access worries about mathematical objects and holes I advocated in Chapter 15. The worry is: why couldn't one give a similar answer to access worries about knowledge of fairies, morality or god? Wouldn't any principled metasemantic story that lets us dissolve access worries suggested by (certain aspects of) our knowledge of mathematical objects and holes in the way I suggested prove too much?

16.1   A Theory of Stipulative (Re)Definition

16.1.1   A Strategy for Stating Weak QV Theses

Linnebo (2018a) and Studd (2019) each assert a kind of Weak Quantifier Variance thesis using the interpretational possibility operator. They say that "there's no maximal quantifier sense" in the sense that however many objects (e.g., sets) one is currently talking in terms of, one could think or talk in terms of more sets. Principles like the Modal Powerset (Axiom 8.11) let me say something structurally similar.[1] However, on their own, such logical possibility claims don't tell us anything about the possibility of *language change*.

So, a natural question is, can one explicitly state a Weak Quantifier Variance thesis that can do the kind of philosophical work proposed in Chapter 15 (i.e., solve certain access problems) without invoking an interpretational possibility operator – or other new conceptual resources? Can I take the scare-quotes off my claims about languages

---

[1] Note how the latter axiom tells us that (it's logically necessary that) it would be logically possible for the total universe to be strictly larger than it actually is.

that "talk in terms of more objects than our own" without paradox (or introducing some such new notions)?

I'll argue for an affirmative answer by providing a theory about the effects of certain acts of **ontologically inflationary stipulative definition**, whereby (informally speaking) a person attempts to start talking in terms of extra objects.

Specifically, I'll consider cases where speakers of a certain kind of language L_0 shift to speaking a different language L_1, by making a certain kind of stipulative definition. I'll propose a concrete "translation procedure" by which sentences in L_1 can be translated into sentences of L_0 which have the same possible world truth conditions (and grounding in fundamentalia).

This translation algorithm serves a few purposes. First, it explains how a person could (in principle) easily go from competent use and understanding of L_0 to corresponding use and understanding of L_1. They could do this by systematically translating sentences of L_1 back to sentences of L_0. Compare the way that an English speaker who knew how to systematically translate German sentences to English ones could be said to understand German.

Of course, if we wanted to fully formalize this claim in a mathematically precise way, we'd need to ascend to a metalanguage which contains a truth predicate for L_0, so we could assert that each sentence in L_1 is true iff its translation in L_0 is true. However, this in no way suggests that you need to have a truth predicate for your own language to shift languages by making a stipulation. Rather, it suffices to have the disposition to assert sentences in the new language just when you'd be willing to assert their translations in the original language. And nothing about having that disposition requires having a truth predicate for your initial language.[2]

16.1.2   Stipulative (Re)Definitions and Their Effects

So now it's time to actually state the relevant (limited) theory of stipulative (re)definition.[3] First, I'll propose a framework for thinking about (idealized) explicit acts of attempted stipulative redefinition. Then I'll make a proposal about their effect on sentences' truth conditions and grounding facts.

For expository simplicity, in this chapter I'll only attempt to describe the effects of stipulative definitions that redefine/re-purpose terms already in L_0 (rather than introducing any entirely new predicates or relation symbols to the lexicon). However, my

---

[2] Now admittedly, familiar issues about general truth predicates and the Liar paradox do prevent us from making certain fully general claims. The range of possible stipulations whose effect on speakers of L_0 I'll be describing won't include any that increase expressive power sufficiently to get one to start speaking current metalanguage L (or any other language that has a truth in L_0 predicate). But I will argue that we don't need any such general claim later in the chapter.

[3] I make an earlier version of this proposal in Berry (2015).

proposal can be easily generalized to capture the effects of (many) stipulative definitions that do introduce new vocabulary.[4]

I propose that we can think of acts of attempted explicit stipulative definition as involving at least two elements. First, there is a sentence S whose (metaphysically necessary) truth the stipulation attempts to secure. In the case of stipulations introducing new kinds of mathematical objects, this sentence S would specify how the mathematical objects being introduced are to be related to one another (and perhaps also how these objects are to relate to various previously understood mathematical objects).

So, for example, a stipulation introducing complex numbers might attempt to secure the truth of some sentence S that conjoins the claim that every pair of real numbers r_1, r_2 corresponds to a complex number $r_1 + ir_2$ (with $r_1 + 0i = r_1$) with the rules for complex multiplication and addition.[5] This stipulation might also fix that the complex numbers are not to be identical to any physical objects, people, etc.

A stipulation attempting to introduce a notion like "city" might (among other things) specify how facts about the application of "city" are to definably supervene on facts about the application of antecedently understood terms/concepts (e.g., "person," "lives in," "spatial point").

Second, there's a fact about which terms in our current language a given act of stipulation is empowered to change vs. which terms' current meaning (or at least pattern of application) are supposed to be held fixed. Note that not all stipulations are made with the intent of introducing new objects. Sometimes we merely wish to change the meaning of a term (or introduce a new term). For example, one might introduce a term like "bachelor" by an act of stipulative definition which puts forward the sentence "$\forall x(\text{bachelor}(x) \leftrightarrow [\text{man}(x) \& \neg\text{married}(x)])$" together with permission to modify the application of the relation symbol "bachelor()" but not "married()," "man()" or any logical vocabulary.

Third, we might want to insist that a stipulative definition is required to keep certain sentences (typically ones that strike us as analytic or conceptually central) in our previous language expressing metaphysically necessary truths. So, for example, if

---

[4] One can use the same tools to model (typical cases of) stipulative definitions that introduce entirely new vocabulary. Modeling the effect of introducing new vocabulary would be straightforward if we were allowed to introduce new (otherwise meaningless) predicates inside the conditional logical possibility operator. However, if our current language contains sufficiently many relations $R_1, \ldots, R_n$ of suitable arity whose application definably supervenes on that of other relations $R'_1, \ldots, R'_n$ we can achieve the same effect with a little ingenuity (by considering situations where the new relations $R_1, \ldots, R_i$ code up the intended application of both $R_1, \ldots, R_n$ and the new relations $N_1, \ldots, N_m$ being introduced).

Similarly, we can model the effect of typical stipulations that use the current application of a term which is being redefined to help specify its new extension (c.f. Linnebo's notion of dynamic abstraction in Linnebo (2018a) discussed in Section 5.4). This will be quite useful. For example, suppose I want to start talking and thinking in terms of (something like) cities. Then I will probably want to expand the application of "is located at" so that cities can count as located in certain regions. However, in doing this, I will not want to allow any change to (the structure of) how "is located at" relates people, physical objects, etc. to spatial points.

[5] Here I take these addition and multiplication rules to be written in a way that implies that complex numbers without an imaginary part are identical to the corresponding real number, as expected.

we start talking in terms of holes and shadows, we might want to preserve the truth of some claim like "for any way of choosing some physical objects, there's a set which collects exactly these objects."[6] However, we can achieve this by taking the sentence S our stipulation attempts to make true, to include these all these conceptually central necessary truths in our old language as conjuncts.

To discuss these issues more formally, let us consider a logically regimented[7] (interpreted) language L_0. I have suggested that one can think of acts of stipulative redefinition as determining an ordered triple as follows:[8]

- A sentence S whose truth is being stipulated (this may include various conceptually central truths which the stipulation is intended to preserve).
- A set of atomic relation-symbols $R_1, \ldots, R_m$ (structural facts about) whose application this act of implicit definition is committed to keeping fixed.[9]
- A specification of whether the stipulation is permitted to change the meaning of $\exists$ and $\forall$ (e.g., 0 or a 1 coding this fact).

Clearly not all such attempted stipulations can succeed:

Definition 16.1 An act of attempted stipulation $\langle S, R_1, \ldots, R_n, 1 \rangle$ made in language L_0 is viable if and only if $\Diamond_{R_1, \ldots, R_n} S$ expresses a metaphysically necessary truth in L_0.

Intuitively, the idea here is that if a stipulation is to succeed, then for each metaphysically possible world, it must be logically possible for the sentence S it attempts to stipulate to be true at that world, while holding fixed (structural facts about) how the relations $R_1, \ldots, R_n$ apply.[10] For example, a stipulation that attempted to secure the truth of a logical contradiction like "$(\exists x)(\text{bachelor}(x) \wedge \neg\text{bachelor}(x))$" would not be viable. Neither would a stipulation that attempted to secure the truth of "$(\forall x)(\forall y)(x = y)$," while holding fixed the current structural pattern of application of "cat."[11] Metaphorically speaking, the fact that a stipulation is viable ensures that it's possible to change how our language carves each metaphysically possible world up into objects in such a way as to make our stipulated sentence S come out true at all possible worlds (while not changing the application of any relations $R_1, \ldots, R_n$ whose applications at these worlds are supposed to be held fixed).

As Definition 16.1 makes clear, in principle whether an attempted stipulative definition is viable can depend on metaphysical possibility facts. However, we can

---

[6] That is, it's logically necessary (fixing the facts about the sets and $P_1, \ldots, P_n$ a list of all the kinds of non-set objects we are currently talking in terms of, as per the next paragraph), that if some otherwise unused predicate P applies to only non-sets, there's a set that collects all and only the objects that P applies to.

[7] I assume that the only logical operators in L_0 are the truth-functional connectives, quantifiers from first-order logic and the logical possibility operators.

[8] Here I will only consider stipulations made in and using interpreted versions of the language of logical possibility.

[9] Note I don't make any assumption that this amounts to keeping the *meaning* of these terms fixed. C.f. the discussion of rich meanings in Section 5.6.2.2.

[10] That is, it's metaphysically necessary that one can do this while holding fixed (the structural facts about) the application of all terms which are not being stipulatively redefined.

[11] For there are metaphysically possible worlds where there are two distinct cats, so it's not logically possible, holding fixed the facts about how "cat" applies that $(\forall x)(\forall y)(x = y)$.

very often recognize a stipulation is viable by a priori logical reasoning alone.[12] In this case I will say that the relevant stipulation is logically viable. That is, a stipulation $\langle S, R_1, \ldots, R_n, 1 \rangle$ is **logically viable** iff $\Diamond_{R_1, \ldots, R_n} S$ comes out true in L_0.

Now I propose a theory about the kinds of acts of attempted explicit stipulative definition considered in this chapter. Making a viable stipulative definition $S, R_1, \ldots, R_n, 1$ while speaking L_0 would shift one to speaking a different language L_1 where the following conditional holds:

The sentence "ϕ" expresses a truth in L_1 if the sentence "$\Box_{R_1, \ldots, R_n}(S \rightarrow \phi)$" is true in L_0

Furthermore, this translation preserves (coarse grained aka possible worlds) truth conditions.[13] We might also say that it preserves truth making or grounding facts, in the sense that whatever suffices to ground the fact expressed by "$\Box_{R_1, \ldots, R_n}(S \rightarrow \phi)$" in L_0 would suffice to ground the truth of the fact expressed by ϕ in I in L_1.

If we make certain further assumptions – that our stipulation is particularly explicit and complete and that our original language L_0 only talks about objects falling under a certain list of kind terms – we can turn the above conditional into a biconditional and provide the promised translation strategy (for stipulations of this kind).

So, first suppose our original language L_0 contains a list of atomic predicates $P_1(x), P_2(x), \ldots, P_n$, which behave like an exhaustive list of kind terms, in the following sense: "$\forall x(P_1(x) \lor P_2(x) \ldots P_n(x))$" expresses a metaphysically necessary truth in L_0. One way for this assumption to be satisfied would be for the relevant language L_0 to include a certain list of kind terms, such that (necessarily) all the objects it carves the world up into belong to one of these kinds.[14]

Now I will say a stipulation of the form $\langle S, R_1, \ldots, R_n, 1 \rangle$ is **categorical** when two conditions are satisfied. First, it attempts to secure the truth of a sentence S which categorically describes[15] the structure of whatever relations are not being held fixed over the relations $R_1, \ldots, R_n$ that are being held fixed. This condition ensures that a categorical stipulative definition pins down, for each metaphysically possible world, a precise pattern for how all these relations $N_1, \ldots, N_m$ (e.g., number, successor, city, hole) are to apply – given the pattern of how the relations $R_1, \ldots, R_n$ apply at that world.[16]

Second, the sentence S being stipulated must have a conjunct asserting that all objects are related by at least one of the finitely many relations in L_1.[17] Thus, viable stipulations will attempt to take us from L_0 to a new language L_1, which also talks in

---

[12] That is, there's no metaphysically possible way the old relations $R_1, \ldots, R_n$ could apply that would prevent S from being satisfied, because it's logically impossible that $R_1, \ldots, R_n$ could apply that would prevent S from being satisfied.

[13] That is, it's metaphysical necessary that the proposition expressed by ϕ in L_1 is true if the proposition expressed by "$\Box_{R_1, \ldots, R_n}(FOO \rightarrow \phi)$" L_0 is true.

[14] Note that this assumption is trivially satisfied for any language like ours which has a predicate expressing a concept "$object_{2020}$" which applies to all objects belonging to kinds I'm currently talking in terms of.

[15] Recall the definition of categorical over in Section 12.1.

[16] Such a stipulation might say that no number is a person, every number is a mathematical object, etc., as well as just stipulating that $PA_\Diamond$.

[17] I restrict myself to considering stipulations which satisfy the latter two conditions purely for simplicity. Ultimately one would want to provide a fuller theory of the effects of stipulative (re)definition that described the effects of stipulative definitions which are not categorical in this sense.

terms of objects falling under some finite list of kind terms $P_1(x), P_2(x), \ldots, P_l(x)$ (though the list of kind terms for L_1 may be different from those for L_0).

Putting this together we have the following condition:

Definition 16.2. A stipulation $\langle S, R_1, \ldots, R_n, 1 \rangle$ made in language L_0 (satisfying the assumption about an exhaustive list of kind terms mentioned in this section) is **categorical** if and only if:

- $\Diamond_{R_1, \ldots, R_n} S$ expresses a metaphysically necessary truth in L_0;
- the sentence S being stipulated is a categorical description of the intended application all relations $N_1, \ldots, N_m$ being stipulatively redefined over the relations $R_1, \ldots, R_n$ being held fixed;
- the sentence S being stipulated includes a conjunct of the form $(\forall x)(P_1(x) \vee P_2(x) \ldots P_m(x))$.

I propose the following rule for translating sentences in L_1 back into sentences in I_0. For any viable stipulation $\langle S, R_1, \ldots, R_n, 1 \rangle$ made by speakers of L_0 will take us to a language L_0 with the following property:

The sentence "ϕ" expresses a truth in L_1 if and only if the sentence "$\Box_{R_1, \ldots, R_n}(S \rightarrow \phi)$" is true in L_0.

In this way speakers of L_0 can translate sentences of L_0 into sentences of L_1. We can also describe the effect of L_0-speakers making such stipulations while working in a metalanguage. Note that when doing this we take ourselves to have truth predicate for L_0 and (in effect) each language you would get by starting in L_0 and making a viable stipulation, but a speaker need not have a truth predicate to make a stipulation and start speaking a new language. A truth predicate is only needed (as it is for anyone) to describe truth conditions in a language.[18]

Note that, unlike some set-theoretic approaches to neo-Carnapian language change (Chalmers 2009; Warren 2014) that describe truth conditions for variant languages by set-theoretically modeling them, the story I've told here can describe truth conditions for languages L_1 that talk in terms of more objects than either L_0 or our current metalanguage. Thus, this technique lets us state a Weak Quantifier Variance thesis without paradox.

16.1.3   Access Worries

In Chapter 15 we wanted to use Weak Quantifier Variance to solve certain access worries. Can the kind of limited claims about language change in this chapter suffice to

---

[18] You might worry that, by analogy with the fact that ZFC can define truth in $\mathbb{N}$, by stipulatively introducing some new objects (perhaps a hierarchy of sets V_α up to some ordinal I can define in the language of conditional logical possibility) I somehow must increase the power of my language so drastically that I could define a truth for my old language in terms of my new language and thus, via my translation, a definition of truth in L_0 of itself. However, this can't be so. If there were some stipulation of the kind considered in this chapter that (while keeping the natural number vocabulary fixed) let one define a truth predicate for L_0 then we could paradoxically define a formula in L_0 itself that defines truth in L_0.

do this? Specifically, can I answer/dissolve access worries about knowledge that I have by considering how speakers of a language L_0 could reliably form true beliefs by making stipulative definitions?

Independent work on access worries (Berry 2020a) suggests that I can. Recall that access worries can be seen as involving a kind of coincidence avoidance reasoning and asking a "how possibly?" question. A realist about some domain (mathematics, morals, etc.) faces an access problem when they seem committed to positing "extra" inexplicable coincidences, that could be avoided by switching to a comparably attractive less realist rival view. Specifically, there seems to be a kind of match between facts about us (e.g., human psychology) and facts about the domain of knowledge in question (as the realist understands it) of a kind that intuitively call out for explanation. Satisfactory explanation for this match[19] seems inconceivable. And we can think of the access worrier as pressing the realist with the following "how possibly" question: how could we (possibly) have gotten the kind of knowledge you take us to have other than by a lucky coincidence?

A common idea from the literature on "how possibly" questions, is that we can solve them by providing a certain kind of toy model. This model must include all the facts about the actual state of affairs that make the fact whose possibility is to be explained *seem* impossible but can simplify and idealize away from other aspects of reality.

Thus, we can plausibly answer access worries by providing a kind of toy model for what seems inconceivable: how someone could have acquired the knowledge we take ourselves to have (as the realist understands it) without benefiting from some spooky coincidence (which the less realist philosopher posing this access worry could avoid). Because of the point in the prior paragraph, the relevant model doesn't need to involve someone speaking exactly our language or having exactly our knowledge. It just needs to explain how someone could wind up with knowledge that's similar to ours in all the ways that seem to raise an access worry.

In the previous subsection I have proposed a tool for creating such toy models. I have tried to provide an attractive picture of how sentences in new languages can systematically inherit truth (and perhaps grounding) conditions from sentences in old languages (and how speakers could go from understanding the original language to the new one) which doesn't require viewing all quantifier senses as mere restrictions of some pre-existing maximal quantifier sense.

I submit that merely considering such a model has the power to dispel a bad philosophical picture, which motivates rejection of the neo-Carnapian explanations advocated in Chapter 15. We dispel a bad picture on which reliably forming true beliefs by making ontologically inflationary stipulative definitions is impossible or implausible, and knowledge of which objects exist is correspondingly much more difficult to explain than knowledge of how predicates apply.[20]

---

[19] By "satisfactory explanation" I mean explanation that banishes apparent realist commitment to an extra coincidence.

[20] For example, a priori knowledge of which logically coherent pure mathematical objects exist/minimum hole steepness seems far more mysterious than knowledge of shades of red that qualify as pink (and similarly knowledge which everyone agrees can be attractively metasemantically explained).

If we can do this then, dialectically speaking, we don't need to make any claim about all possible languages, or give non-trivial meaning to claims about an "absolutely general quantifier sense" (even to deny that there is one) in order to do the work we want accepting weak quantifier variance to do.[21]

16.2   When Are Meta-semantic Explanations Plausible?

This brings us to the second worry I want to consider in this chapter. One might also object to the Quantifier Variance thesis as follows. If our use can change the meaning of our quantifiers (allowing for a metasemantic answer to access worries about our knowledge of minimum hole indentation), then why doesn't fairy believers' talk of fairies change the meaning of our quantifiers so "there are fairies" is true?

I first want to note that this worry is merely a special case of a problem that everyone faces. Even philosophers who reject quantifier variance will allow that we can metasemantically account for (i.e., dispel access worries about) our access to some facts about *how properties apply*. But the same kind of reasoning used to raise the worry would equally well call into question uncontroversial metasemantic explanations. For example, presumably, the correct explanation of our accuracy regarding what colors qualify as pink (in the sense relevant to banishing access worries) involves the following fact: if we were inclined to differentiate colors in another way then the meaning of "pink" would have been different. (I take the acceptability of such metasemantic explanations to be fairly uncontroversial in cases where no quantifier variance is required to deploy them.) But one could just as well ask: if our use of "red" can change what predicate this word expresses then why doesn't antivaxxers' use of "autism-causing" change the meaning of that predicate, so their beliefs come out true?

For this reason, I think the problem of providing a principled theory of when access worries can be metasemantically answered is a problem for everyone. To my knowledge, no complete and satisfying story about how use determines meaning in either case has yet been developed. But it seems to me that essentially the same tools (appeal to a distinction between more and less definitional/would be analytic aspects of our use, appeal to more vs. less joint-carving/natural kinds ways of conceptualizing the world in the Siderian sense discussed in Section 11.4.2) seem available in both cases.

---

[21] Note that the neo-Carnapian has no problem allowing that we can (and do) quantify over everything.

   Also, one might object to the above argument as follows. The neo-Carnapian is willing to actually make ontologically inflationary stipulative definitions. They in effect take themselves to have a reliable faculty of knowledge gain/preservation where (making ontologically inflationary stipulative definition and leads to knowing that the stipulated sentence is true, and that certain sentences from ones old language must continue to express truths). Doesn't this commit us to thinking something about our own language, not any toy model?

   Briefly put, my answer to the latter challenge is this. Merely accepting neo-Carnapianism/quantifier variance (and/or using the interpretational possibility operator to state it) can't create any new problems with the Liar in this way. Everyone is committed to our being able to gain/preserve knowledge via *some* faculty of stipulative definition. *Everyone* acts as though we have a faculty of reliably accepting true sentences by making *categorical* stipulative (re) definitions like "For all x, x is a bachelor iff x is an unmarried man."

Despite this dialectical point, I will close this chapter by proposing a tentative account of when access worries can be metasemantically answered. I do this because I think it provides a nice case study for the usefulness of the logical possibility operator outside philosophy of math and logic. However, it should be noted that my aim in these sections is to more to advocate a research program than to make a final proposal. None of my claims or arguments elsewhere in this book will depend on accepting the particular proposal I'll make in the next section.

16.3   Evaluating Metasemantic Answers to Access Worries

Roughly speaking, I propose that a philosopher can solve an access worry metasemantically when the bundle of good epistemic states (including but not limited to having certain true beliefs) they claim to have that generates this access worry can be rationally reconstructed in a certain way. They can answer access worries by showing that the controversial knowledge and other epistemic good statuses they claim to have could be gotten via a certain kind of process involving stipulative definition – whether or not they ever actually made such a stipulation.

So, for example, Arthur Conan Doyle can metasemantically answer access worries about his claimed knowledge of fairies, if he can produce a certain kind of rational reconstruction. Specifically, we imagine someone starting from an uncontroversial ground language body of knowledge and epistemic faculties.[22] And we ask whether this person could acquire Doyle's total collection of controversial claimed epistemic good statuses via making stipulative definitions (and perhaps deploying some of their other uncontroversial good faculties).

But what do I mean by rationally reconstructing someone's controversial epistemically good status

Crucially (as I will understand the term), it isn't enough to rationally reconstruct someone's (relevant controversial) epistemic status to tell a story about how they could have come to rationally accept that certain *sentences* express true propositions. We must tell a story which simultaneously rationally reconstructs a number of other things, like their dispositions to reach certain conclusions in response to observations or to apply induction in certain ways. That is, we can't answer access worries about Conan Doyle's fairy knowledge merely by noting that someone started from uncontroversial assumptions could have made a series of viable stipulative definitions which would ensure that all the *sentences* involving the world "fairies" Doyle eccentrically accepts would express truths (in their personal idiolect). We also need to justify (enough of) Doyle's dispositions to infer between accepting claims about "fairies" and observations of fluttering lights in gardens, etc. Additionally, our reconstruction must allow for

---

[22] By uncontroversial knowledge, faculties, etc. here I mean knowledge which the person pressing an access worry doesn't dispute. The initial state I invoke here is simply that of a person whose language faculties and knowledge all parties agree doesn't create an access problem (of the kind at issue).

(enough of) Doyle's dispositions to make inductive inferences to be justified.[23] Recall Goodman's point that we take certain predicates but not others ("green" but not "grue") to be projectable (Goodman 1955)) and that stipulatively redefining "green" could diminish or destroy our warrant to make abductive arguments using it.[24]

So, there are two questions we need to answer to fill in the above rational reconstruction framework. First, what kind of claimed epistemically good states do we need to reconstruct? Second, how does making stipulations change these states?

16.3.1   Sketch of Epistemic Dynamics of Stipulative Definition

I'll only give a partial answer to these two questions here. But the key idea behind my proposal is simple. Acts of stipulative definition can (so to speak) both create and destroy. A person who stipulates gains warrant to accept the stipulated sentence S, and knowledge of whatever proposition it expresses in their new language. However, stipulating can also destroy valuable relations to one's old language such as warrant to accept certain sentences that expressed truths in one's old language, deploy certain observation procedures or abductively project certain predicates, etc. Accordingly, we need to keep an eye on these things when evaluating rational reconstructions that attempt to answer access worries.

Consider someone who makes (what they know to be) a viable stipulative redefinition $\langle S, R_1, \ldots, R_n, 1 \rangle$ of the form modeled in Section G.1.2 in the online appendix. What effects does this have? Most obviously (as we said), the stipulator can gain knowledge of S, the sentence whose truth the stipulation attempts to secure. Plausibly they can get warrant for accepting S[25] comparable to their warrant for thinking this stipulation was viable.

More generally, after making a categorical viable stipulation $\langle S, R_1, \ldots, R_n, 1 \rangle$, (roughly speaking) someone will have[26] at least as much warrant to accept a sentence ϕ in their new language L_1 as they had to accept both the claim that the stipulation was viable and the *translation* of ϕ back into their ground language L_0.

Thus, someone who makes a categorical stipulative definition of the kind I've modeled will be able to carry over some (but not all) of their dispositions to accept (and reason to or from) sentences in their old language.[27]

---

[23] This requirement blocks tricks which redefine "fairies" into existence while justifying the preservation of ordinary observational practices where, e.g., meaning changes where "fairy" is defined to apply to one kind of thing within my future light cone and another kind of thing outside of it (where you are guaranteed not to observe it).

[24] Within the Siderian framework of Section 11.4.2, all these claims are associated with taking notions to be metaphysically joint-carving.

[25] That is, they can get justification for believing what it expresses in your post-stipulation language.

[26] See the important caveat in the discussion of Williamson on analyticity in Section 16.4.

[27] I say "at least" (rather than just roughly) for the following reason. Consider someone who makes a plausible but dicey (i.e., not clearly viable) stipulation attempting to secure the truth of some sentence S. Perhaps, for example, they attempt to introduce a pure mathematical structure satisfying axioms that are logically coherent, but whose logical coherence they are only justified in being moderately confident in. They should be somewhat cautious about whether the stipulation succeeded or failed. Thus, they will only have modest justification for accepting S. But a natural thought is that (though alternative accounts are surely possible) if

Obviously, stipulations that redefine our terms or quantifiers can change the truth-value of sentences. So we can't always homophonically translate between our initial language and the language we speak after making a stipulation. However, we usually expect stipulations to preserve homophonic translation for most sentences in our language. My story accounts for this by allowing us to specify a list of relations whose application is to be held fixed. It is easy to see that any sentence which is content-restricted to the fixed relations (in the sense defined in Chapter 7) has the same truth-value in both languages.[28]

Also note that (intuitively and in terms of the picture I've provided here) stipulations have an effect on things we do with words besides accepting sentences. It can disrupt our warrant for using observation procedures, abduction and homophonic translation. For example, if I stipulatively redefine "more massive than," I can't assume that:

- old observation procedures for applying "x is more massive than y" will remain reliable;
- syntactically simple predicates stated using this phrase will continue to express something projectable/abduction friendly;
- I can continue to translate certain people's talk of "massiveness" (my past self, and people she would have translated homophonically) homophonically.

Overall, we see to face a kind of trade off when considering what kinds of powers to give the acts of stipulative definitions in our rational reconstruction. The more words' applications a stipulative definition is empowered to change, the easier it is for that stipulation to succeed (and for a person to have warrant for assuming that stipulation is viable).[29] On the other hand, the more words' applications a stipulation is empowered to change, the greater change it can make in what observation procedures, inductive inferences, etc. are warranted (and thus the more potential problems it raises for rationally reconstructing the aspects of someone's overall epistemic state that generate access worries).

16.3.2   Examples

In the case outlined in Section 16.2, Arthur Conan Doyle couldn't explain the total epistemic state he (presumably incorrectly) takes himself to be in, by some act of

a stipulation fails, then L_1 will just be L_0. So they should continue to be very confident in simple logical truths like, "If a ball is red and round then it is red" (much more confident than they are in the viability of their attempted stipulation), as these sentences express truths in both languages and are hence guaranteed to express truths whether or not the stipulation succeeds.

[28] For example, (basic logical possibility reasoning involving my stipulation makes clear that) viable stipulative definitions that aren't empowered to change the application of the terms "person" and "child of," can't change the truth-value of "Some person has more than one child." As few sentences we use in daily life truly quantify over all objects, as long as we make stipulations that keep fixed the vast majority of our vocabulary, we will be able to homophonically translate the vast majority of sentences from our old language.

[29] For example, it might be unclear whether you could make all the sentences of some formalization of an essay by Leibnitz come out true with a stipulation only empowered to change quantifier meanings and the application of the term "substance" but obvious that one could do so if one were allowed to change the meaning of all non-logical vocabulary.

stipulative definition (together with deployment of faculties not currently being called into question), for the following reason.

Imagine a person who starts out with an uncontroversial good epistemic state. Such a person could, certainly, come to rationally accept all the *sentences* of Doyle's controversial theory about fairies, by making a suitably empowered act of stipulative definition. For an act of stipulative redefinition that was empowered to change the application of sufficiently many terms in the theory ("wing", "fly", "fairy", "see", etc.) could easily be viable, and could easily be known to be viable the protagonist of our thought experiment.

However, someone who did this would lack many epistemically valuable relationships to their own language which Doyle takes himself to have:

- They wouldn't be justified in treating the observation procedures they used to associate with terms like "wing" that have been stipulatively re-defined as continuing to be reliable.
- And they (*ceteris paribus*) wouldn't be justified in translating sentences uttered by certain kinds of people who don't accept key tenants in Doyle's theory of fairies as meaning something that's true if and only if Doyle's orthographically identical sentence would be true.
- They would have no reason to think their new terms would continue to express concepts that are particularly abduction friendly, in the way that green is but (famously) grue isn't. So they wouldn't be justified in making the abductive inferences about wings and seeing etc. Doyle does.

As Conan Doyle takes himself to be justified in doing all these things, he cannot rationally reconstruct the controversial parts of his own situation in this way.

Contrast this with the metasemantic answer to access worries about our knowledge of which logically coherent pure mathematical structures exist advocated in Chapter 15.[30] Philosophers plausibly can rationally reconstruct the kind of knowledge and epistemically valuable relationships to language they take us to have.

Consider someone who started out speaking a language L_0, that didn't talk in terms of anything like mathematical objects (but had good methods of reasoning about logical possibility). They could gain knowledge that some axiomatic principles like $PA_\Diamond$ are true by making logically viable stipulative definitions and then plausibly acquire all the other epistemically valuable statuses we take ourselves to have via processes that are taken for granted by all parties pressing this access problem:

- We don't tend to have observation procedures associated with mathematical objects (except perhaps for counting procedures associated with the natural numbers).[31]

---

[30] Note that I take solving this access problem to be the only part of what's required to answer general access worries about our access to mathematical objects. The harder part is accounting for our knowledge of logical possibility facts. See Berry (2018b) for proposals regarding that.

[31] I suspect that these can be rationally reconstructed from logical possibility reasoning in a broadly Fregean manner, but I won't say more about that here.

- As regards homophonic translation dispositions and reference magnetism, contemporary mathematicians don't seem to take their choice of which logically coherent pure mathematical structures to talk in terms of to be strongly reference magnetic. They interpret communities that (seem to) assert the existence of different kinds of pure mathematical objects as speaking truly nonetheless, rather than translating them homophonically as disagreeing with each other and us.
- The stipulation wouldn't be empowered to change the meaning of "wing" and so forth, so they wouldn't have any problem preserving permission to treat these notions as joint-carving. As for the terms which they are redefining (those playing the role of number and successor), they could then gain knowledge that these notions are joint-carving and abduction friendly (e.g., useful for stating physical and mathematical laws) to the extent that we think they are, by noting how simple a description these properties and mathematical structures have in terms of conditional logical possibility.[32]

Accordingly, I think this kind of rational-reconstruction-based framework is worth developing. But, all this is just a preliminary sketch. I'll end with two areas where I think more work is needed.

The first area concerns how far removed our rational reconstruction can be from what actually occurred. Presumably, rational reconstructions shouldn't involve long inference chains which that bear very little relation to anything the person being rationally reconstructed actually did. For example, imagine a normal person with an ordinary lifespan and no access to computers who claims to be able to know, for all strings of 50 numbers, whether that exact string occurs in the first billion places of the decimal expansion of π. This claim would generate access worries that couldn't be answered by pointing out that one could in principle get this knowledge by deploying ordinary mathematical abilities over a long time.[33]

Secondly, some kinds of metasemantic explanations might require stipulations directly phrased in terms of observation procedures. For instance, consider stipulative definitions of (phenomenological) color terms, or the hole example in Section 15.1.2. Adequately rationally reconstructing the kind of knowledge at issue here might require appeal to stipulative definitions that attempt to secure the reliability of observation procedures rather than the truth of sentences. I think my story about stipulative definitions could plausibly be generalized to account for such knowledge (though see Berry (2019a) for a paradox about notions of solidity that arises on route to doing so). But filling in all that is a task for another book.

---

[32] A different way to think of this point is: they could gain justification for treating certain hypotheses in their new language as elegant, a priori attractive, and supportable by induction via the fact that they had warrant for finding the translation of these hypotheses in their old language similarly attractive. Note that saying this doesn't require us to take there to be principled facts about whether the objects or predicates a person talks in terms of after neo-Carnapian language change are literally the same as those they talked in terms of prior to this language change. C.f. Section 5.6.2.2.

[33] Note: I take this to be a general issue for everyone who accepts the idea of rational reconstruction, not something that depends on any specific worries I've raised here.

16.4   Contrast with Other Answers to This Challenge

Let me conclude by discussing three ways this approach to evaluating metasemantic answers to access worries promises to be helpful.

First neo-Carnapians (myself included) sometimes informally say that one can metasemantically explain our knowledge of which mathematical structures exist because all logically coherent choices of mathematical structures are "equally good."[34] I think such remarks are evocative and helpful. But this isn't exactly true (if understood literally and straightforwardly). For surely the mathematical structures mathematicians posit and use *are* especially good in various ways: with regard to beauty, letting one state or prove general principles that illuminatingly explain other mathematical phenomena and the like. Indeed, these structures might well be especially joint-carving in something like Sider's sense.

The rational reconstruction framework advocated in this chapter lets us avoid this awkwardness by cashing out arguments for and against metasemantic explicability without appealing to this obscure notion of being "equally good."

Second, it is sometimes suggested[35] that the mere fact that pure mathematical theories "conservatively extend" our non-mathematical talk in the way Field (1980) defines,[36] explains why we can give a Quantifier Variance explanation for our knowledge of pure math theories (given knowledge of logical coherence).

But this can't be a complete explanation as it stands. In general, showing that your beliefs about *F*s conservatively extend your beliefs about non-*F*s is not enough to dispel access worries about your knowledge of *F*s. For example, noting that your moral beliefs conservatively extend your physical beliefs (in the sense of not letting you prove any new claims stated in purely physicalist language) is not sufficient to show that no access worries arise with regard to moral knowledge – pace certain readings of Scanlon (2014).

Third, other popular neo-Carnapian theories like Thomasson's influential (Thomasson 2015) appeal to (epistemically) analytic truths, while my proposal avoids controversial commitment. As I'll now clarify, it lets us metasemantically explain our knowledge of certain facts without any suggestion that the relevant facts are analytic or impossible for anyone who possesses the relevant concept to rationally doubt.

This lets us better accommodate Quine's thought that, "The lore of our fathers is . . . a pale grey lore, black with fact and white with convention" (Quine 1960) and Williamson's more recent arguments against there being (in any interesting sense) conceptual truths (Williamson 2008). Williamson argues that although claims like "vixens are foxes" are conceptually central, even these claims aren't epistemically analytic or unreasonable to doubt. He points out that philosophers *do* seem to be able to

---

[34] See Clarke-Doane (2020) and Berry (2018b).   [35] One might read Rayo (2015) as saying this.

[36] Introduce a one-place predicate "$M(x_1)$" meaning intuitively x_1 is a mathematical object. For any assertion A, let A^* be the assertion that results from restricting each quantifier of A with the formula "not $M(x_1)$." [Then our mathematical theories seem to "obviously" satisfy the follow conservativity condition:] Let A be any nominalistically stateable assertion. Then A^* isn't a consequence of S unless it is logically true (Field 1980).

intelligibly and rationally doubt even the most basic and conceptually central principles and inference rules for philosophical reasons (e.g., McGee has seemingly argued against *modus ponens* being exceptionlessly a legitimate inference). And he argues that even the most conceptually central principles associated with a term (in the sense of constraining our interpretation of that term) principles are dubitable for the following reason. Even the most seemingly conceptually core principles associated with a given term can fail to express truths if they fail to harmonize with one another in the way that the inferences associated with the spurious concept "tonk" do.[37] Indeed (as he further suggests) maybe our naive concept of truth is an example of a natural language concept whose core inference rules fail to harmonize with one another in this way.[38] Thus, even in cases where one's principles are actually harmonious, it can be rational to entertain some doubts corresponding to the possibility that these principles aren't harmonious.

To see why the proposal I've sketched doesn't require appeal to indubitable conceptual truths or (epistemic or metaphysical[39]) analyticities, first consider cases where an explicit stipulative definition has been made.

The theory of stipulative definitions in Section 16.1.2 doesn't imply (and I would positively deny) that acts of successful stipulative definition yield knowledge that's completely certain and indubitable. For one thing (in line with Williamson's point that we can rationally worry that our concepts are tonk-like (Prior 1960)) note that my "epistemic dynamics" say that acts of stipulative definition merely transform warrant for accepting that a certain stipulation is viable into comparable (perhaps *slightly* lesser) justification for accepting the stipulated sentence S.[40] Leaving this room for rational doubt about whether a stipulation is viable (in cases where it actually is viable) reflects exactly the kind of concern Williamson presses in regard to tonk (that things that seem like conceptually central truths involving some notion might not be suitably coherent/harmonious with each other), as they arise in the case where we are

---

[37] Recall that "tonk" (Prior 1960) has the introduction rules for or and the elimination rules for and thus lets you reason from any sentence A in your language to any sentence B.

[38] I think Williamson's other example of such conceptually core truths being rationally dubitable and indeed sometimes false – "phlogiston has role R" – is less satisfactory, because (along the lines of Boghossian) the friend of conceptual truths will presumably say that's what's a conceptual truth is "if there's phlogiston then it plays role R," for obviously people can understand the concepts oxygen and phlogiston and debate which one explains the behavior of fires etc., without having confidence either hypothesized substance plays the relevant role.

[39] It may also be worth noting that, like most contemporary neo-Carnapians I don't take metasemantically explicable knowledge to involve relevant being metaphysically analytic in the sense Boghossian (1996) criticizes. I don't take it to follow from the idea that considering the results of stipulative definition can help solve or reduce certain access worries about our knowledge of which pure mathematical objects exist to suggest that any interesting pure mathematical existence facts are constituted by or grounded in more made true by facts about how humans use language. Indeed, the story about grounding inheritance during acts of stipulative re-definition proposed above suggests quite the opposite conclusion.

[40] Perhaps the appearance that stipulative definition gives rise to knowledge of indubitable claims comes from considering a limited diet of examples. We consider stipulative definitions whose form makes it particularly transparent that they are viable: specifically categorical definition, which define some new term N by stipulating that $(\forall x)(N(x) \leftrightarrow \phi(x))$ where ϕ only contains old vocabulary whose meaning the stipulation is licensed to change.

attempting to introduce a new concept by stipulating some relevant conceptually central principles. What about conditional claims like "If such-and-such a stipulation is viable then *S* is true?" I'm only advocating the theory of stipulative definitions as a truth of philosophy of language, not an indubitable truth. So, I'm not committed to these principles being analytic or indubitable.

The fact that we are merely *reconstructing* our knowledge *as if* we had made certain stipulations rather than actually making them allows for additional possibilities for doubt. And I think some of these additional possibilities are worth highlighting, because they show how the type of metasemantic response to access worries I'm advocating can handle certain apparent problem cases that more traditional and aggressive neo-Carnapian answers to access worries can't.

For example, we might rationally reconstruct someone's knowledge of Turing degrees by considering a suitable abstraction principle involving either Turing machines or partial recursive functions. Someone whose knowledge of Turing degrees is being rationally reconstructed needn't have exactly the knowledge that someone who literally stipulated either one might have. They need not have great confidence that facts about Turing degrees reflect one possible stipulative definition rather than other if evidence suggested that (proofs of the equivalence of the two definitions of computability were some kind of massive illusion and) these notions could actually come apart, they might rather unsure what to say.

Relatedly, a speaker may not know that a particular act of explicit stipulation is enough to reconstruct their epistemic state. Thus, they can (intuitively reasonably) worry they are overlooking some further aspect of their linguistic practice which makes the obviously viable stipulation they have in mind insufficient. For example, the inference dispositions you have regarding bachelorhood aren't transparent. Considering new cases (like the question of whether the Pope is a bachelor) can surprise you by revealing new inference dispositions you didn't realize you had lurking.[41] For this reason, even in cases where the viability of a stipulation seems obvious as a logico-mathematical claim, a natural language speaker whose knowledge can be rationally reconstructed (for the purposes of solving access worries) by considering such a stipulation can reasonably doubt the truth of the sentences stipulated in this reconstruction.

---

[41] You might think when you infer from learning that John is single and an adult man to the conclusion that he's a bachelor nothing would change your mind, and then be surprised to find that learning that John is a religious celibate does change your mind.

17 Logicism and Structuralism

In this chapter, I will briefly discuss how the broader approach to the philosophy of mathematics sketched in the book relates to two big ideas in the philosophy of mathematics: Logicism and Structuralism. I will suggest that my proposals qualify as (at least broadly) Logicist and Structuralist and offer some advantages over certain existing forms of Logicism and Structuralism. However, I'll note that they certainly don't vindicate some of the more ambitious key ideas which have traditionally been associated with Logicism.

17.1 Logicism

17.1.1 Arguably a Kind of Logicism

Let's start with Logicism. In a *Stanford Encyclopedia* article, Tennant (2017) characterizes strong and weak Logicism about a given branch of mathematics in terms of acceptance of the following pair of theses:

- All the objects forming the subject matter of those branches of mathematics are logical objects.
- Logic – in some suitably general and powerful sense that the logicist will have to define – is capable of furnishing definitions of the primitive concepts of these branches of mathematics, allowing one to derive the mathematician's 'first principles' therein as results within Logic itself. (The branch of mathematics in question is thereby said to have been reduced to Logic.)
- All truths (strong logicism) or all theorems (weak logicism) are logical truths. (Tennant 2017)

The Nominalist and Quantifier Variance Realist approaches to the philosophy of mathematics developed here fairly clearly satisfy the first requirement above. We've shown how to categorically describe (or, in the case of set theory, otherwise pin down definite truth-values for sentences about) pure structures using only logical vocabulary in the language of logical possibility.

The Nominalist proposal trivially satisfies the first and second requirements (by not positing any mathematical objects), and clearly satisfies the third (one can prove $\phi \leftrightarrow \phi$, for each nominalistic logical regimentation of a pure mathematical claim).

What about the neo-Carnapian Realism about mathematical objects outside set theory I ultimately advocate in Chapter 15? Admittedly it doesn't satisfy the letter of the second and third requirements above, for my neo-Carnapian Realist takes it to be logically contingent (though metaphysically necessary) that any mathematical objects exist, and hence that various existence claims about mathematical objects are true.

However, I will argue that my favored neo-Carnapian Realist view satisfies the spirit of the requirement that all mathematical objects are logical objects and mathematical truths are logical truths insofar as it endorses the following claims:

- Apparent quantification over the sets is to be understood as making pure claims about logical possibility.
- Knowledge of other mathematical objects (i.e., objects in other mathematical structures which don't give rise to paradox) can be gained by logical knowledge and stipulative definition alone.
- Although mathematical objects are logically contingent, in some sense (which I don't claim to have made precise here) the "job" of pure mathematical objects is to help us explore facts about logical possibility and necessity which *are* logical necessary truths. Facts about mathematical objects are facts about logically necessary truths concerning what structures are logically possible/impossible in something like the way that facts about mountains could be said to be facts about rock distribution.

I'd also suggest that differing from traditional Logicism by saying mathematical objects' existence is merely metaphysically (rather than logically) necessary is an advantage. It lets us agree with contemporary mathematical practice, which takes all set-theoretic models to represent genuine logical possibility. We don't say its; logically necessary that math objects exist, but rather that all the knowledge that we need to reliably come to believe true quantified claims involving mathematical objects is logical knowledge and (something like) knowledge by stipulative definition.[1]

17.1.2   Traditional Benefits of Logicism

I submit that the approaches in the prior section also satisfy some traditional motivations for Logicism. For example, both support Frege's claim that (contra Kant) mathematical truths and knowledge are independent of any intuitions about space and time (Frege 1980). Both also connect mathematical investigation to general subject matter neutral laws in a way that helps explain the widespread usefulness of mathematics. Much more could be said about the various ways the investigation of logical possibility (which I take to be the core project of mathematics) is useful, but I won't say so here.

What about showing that mathematics is analytic? Arguably this is a desideratum which motivates (or has motivated) many traditional forms of Logicism. Suppose we take analytic truths to be those which (in some sense) "follow from logical laws and

---

[1] We also avoid traditional Logicism's famous bad company problems, about how to say, without arbitrariness, which of the various possible internally coherent but jointly incompatible hypotheses describing of existence and identity conditions for supposed mathematical objects (like Hume's principle) are true. The quantifier variantist will say that mathematicians are free to choose any internally coherent (suitably quantifier restricted) posits characterizing pure mathematical objects they like.

definitions alone" (following Frege). Then (at least) the modal Nominalist proposals of Chapter 10, which say mathematical truths all *are* logical truths, clearly imply that mathematics is analytic.[2]

However, showing that all pure mathematical truths technically qualify as analytic winds up doing little of the work philosophers have traditionally associated with this thesis.

For example, it doesn't show that all mathematical truths can be arrived at via deductions from indubitable premises – or even that they are accessible to us at all. Familiar Gödelian considerations still show that many mathematical truths will be cognitively inaccessible to us. Switching from thinking about mathematics as investigating abstract objects, with no special relationship to logic, to thinking of mathematics as investigating abstract, objective, necessary truths about logical possibility and extendibility makes no difference to those arguments.[3]

Accepting that all mathematical truths are analytic for the reasons given here also doesn't immediately vanquish access worries for traditional Platonism. For if it's mysterious how we can get knowledge of platonic objects, it can seem equally (or almost equally) mysterious how we could get knowledge of the powerful and far from cognitively non-trivial logical truths that are identical to, or required to recognize, mathematical truths on the views just mentioned.

This is not to say that adopting the kind of Logicism I've advocated here has no epistemic benefits. All versions of the logical-possibility-centric approach to mathematics I've advocated here *do* reduce the problem of pure mathematical knowledge to a problem of accounting for knowledge of general logical laws that constrain all objects and relations (much as the laws of physical possibility do). And I think making the latter reduction can provide help in dispelling access worries (for reasons we will briefly discuss in Section 19.3).

17.2   Structuralism

Now let's turn to Structuralism. In this section I will review what I take to be the central Structuralist intuitions and note how my proposals largely satisfy them. Insofar as

---

[2] Perhaps versions of the neo-Carnapian Realism about ordinary mathematical objects could be argued to imply the same conclusion (via more substantive assumptions about definitional truth and knowledge by stipulation). Although I personally have tried to show that this view can be developed without endorsing the existence of any analytic truths at all, fans of analytic truths might argue as follows. Suppose I attempt to introduce the natural numbers by a stipulative definition that attempts to secure the truth of $PA_\Diamond$ by the kind of stipulation considered in Section 16.1. Then (one might say, although I don't personally say for the reasons discussed in Section 16.4) that the following conditional is analytic: if this stipulation is viable then $PA_\Diamond$. We can choose parameters for this stipulation such that it's a truth of logic alone, knowable without appeal to any non-logical constraints on the application of vocabulary we are holding fixed, that this stipulation is viable (as per the definition of logical viability in Section G.1). On this view, pure mathematical truths are known by a combination of logical insight and access to something like stipulative definitions.

[3] In both cases, we have reason to think our axioms are recursively enumerable and sufficiently strong to prove a version of PA, and hence incomplete.

much of this book can be seen as philosophically and mathematically developing Hellman's Modal Structuralist development of Putnam's modal perspective on mathematics, this won't be surprising.

In his massively influential "On What the Numbers Could Not Be," Benacerraf (1965) quotes R. M. Martin to the effect that mathematicians are mostly interested in "mathematical structures," their possibility and how they can relate to one another, rather than the nature of the objects which form these structures, and how they relate to other objects (topics which tend to interest philosophers. He writes:

> The attention of the mathematician focuses primarily upon mathematical structure, and his intellectual delight arises (in part) from seeing that a given theory exhibits such and such a structure, from seeing how one structure is "modelled" in another, or in exhibiting some new structure and showing how it relates to previously studied ones . . . But . . . the mathematician is satisfied so long as he has some "entities" or "objects" (or "sets" or "numbers" or "functions" or "spaces" or "points") to work with. He does not inquire into their inner structure or ontological status. (Benacerraf 1965)

Since then, Structuralism, the idea that mathematics is (in some important sense) "the science of structure," has enjoyed great popularity in the philosophy of mathematics and been developed in various ways. I'll highlight how the philosophies of mathematics developed in this book qualify as Structuralist, by considering three enduringly central and popular Structuralist themes in the quote above.

First there's a kind of "mathematicians' freedom" intuition: mathematicians are free to study any logically coherent pure mathematical structure they like[4] – whether this structure is instantiated or not. The Nominalist and quantifier variantist proposals developed in this book clearly vindicate this intuition, by allowing that mathematicians are free to study pure mathematical structures characterized by arbitrary logically coherent axioms.

Second, there's the idea that all possible instantiations of a given mathematical structure (e.g., "the strings on a finite alphabet in lexical order, an infinite sequence of strokes, an infinite sequence of distinct moments of time, and so on") are equally relevant to the mathematical study of that structure. My Nominalist and Quantifier Variance Platonist proposals both reflect this by saying that mathematicians (directly or indirectly, via witnessing mathematical objects whose job is to help us investigate such facts) study pure logical possibility. That is, they study facts about logically possible or necessary simpliciter, what's allowed or precluded by the most general laws of how any objects could be related. From Axiom 8.5 (Importing), note that if it's logically possible that ϕ, then the sentence created by replacing relations in ϕ with other

---

[4] Stuart Shapiro (1997) approvingly quotes Resnik (1981) to similar effect, saying, "Take the case of linguistics. Let us imagine that by using the abstractive process . . . a grammarian arrives at a complex structure which he calls English. Now suppose that it later turns out that the English corpus fails in significant ways to instantiate this pattern, so that many of the claims which our linguist made concerning his structure will be falsified. Derisively, linguists rename the structure Tenglish. Nonetheless, much of our linguist's knowledge about Tenglish qua pattern stands; for he has managed to describe some pattern and to discuss some of its properties. Similarly, I claim that we know much about Euclidean space despite its failure to be instantiated physically."

relations of same arity is also logically possible. Thus, for example, the Nominalist version of a statement about number theory will immediately imply the corresponding claim about each of the instances of the natural numbers structure. In this way the facts about logical possibility studied in, say, number theory constrain the behavior of other instances of the natural number structure (e.g., structures of stroke sequences in which to "the right of" play the role of successor and sequences of clock chimes in which "after" plays the role of successor) without assigning any particular choice of relations metaphysical special status or much conceptual priority.[5]

Third, there's Benacerraf's point that it would be odd to suppose that there are deep or interesting facts about which objects in different mathematical structures are identical to each other, e.g., deep facts about which set (if any) is identical to the number one (given the different ways of reducing number theory to set theory which have been employed by different mathematicians). My Nominalist and Quantifier variantist can both honor this point by saying that such facts reflects something like a stipulative definition (what kind of logically coherent total mathematical universe do we want to talk in terms of?) rather than any non-trivial facts about logical possibility.[6]

17.2.1 Advantages over Ante-Rem Structuralism

Let me close this chapter by suggesting that, like Hellman's Modal Structuralism, the views developed here enjoy some advantages over Stuart Shapiro's *ante rem* Structuralism (Shapiro 1997) (which honors Structuralist intuitions in a very different way).

According to Shapiro's *ante rem* Structuralism there are such abstract platonic objects called structures, and these are what mathematicians study. For example, there is a natural number structure which all ω sequences instantiate. There are other objects called "positions," which belong to a given structure (e.g., the natural number structure will have a countable infinity of different positions belonging to it). And each structure will also include or specify some way for some finite collection of relations to apply to these positions. Roughly speaking, the idea is that the natural number structure abstractly represents what it is that the things we described as instances of the natural number structure (isomorphic collections of strokes, spatial points, etc.) have in common. Furthermore, there's no fact about whether two positions within two such

---

[5] Admittedly, when stating the Nominalist version of a given number theoretic sentence you will pick arbitrary non-mathematical relations of the right arity. However, this choice is clearly superficial as, e.g., the claim you get by picking one choice of non-mathematical vocabulary is obviously and immediate logically equivalent to the claim you get by making any other choice

[6] There might be determinate facts about whether the numbers are identical to various other kinds of mathematical structures (though presumably not to the sets considered in higher set theory, as I've argued that the latter is better explicated potentialistically). But such identity facts between different mathematical structures are not taken to outrun the logical consequences of things we explicitly believe and treat as conceptually central about the relevant mathematical objects and their intended relationship to one another. Like knowledge of which pure mathematical objects (outside higher set theory) exist, our knowledge of identity claims relating objects in distinct mathematical structures is taken to be the kind of knowledge one could get by stipulative definition. For example, these identity facts aren't taken to be reference magnetic, relevant to the statement of abduction friendly natural laws.

different mathematical structures (like the sets and the natural numbers) are distinct. In this way we avoid the dilemma of which sets are identical to the number 1.

Three problems naturally arise for this reifying Structuralism. First, there's an immediate metaphysical oddness to saying that there's no fact about whether positions within distinct structures are identical to one another as a matter of metaphysics not mere ambiguity in which objects the terms number and set denote. Second, there's a much-discussed problem about how to account for the fact that i and $-i$ are distinct if we say (as Shapiro does) that positions in structures *only have* relational features, since (in a sense) i and $-i$ have all the same relational features. Adopting the Nominalist or neo-Carnapian Realist version of Structuralism I've developed in this book (or Hellman's original modal Structuralism) avoids both problems.

Admittedly, on the neo-Carnapian Realism about mathematics I ultimately favor there are special objects, the natural numbers, which mathematicians consider. But these mathematical objects only have a special relationship to number theory in the sense that their "core job" is to help us study logical possibility and necessity facts which equally constrain how all objects can be related by all relations. And all pure mathematical questions can be formulated in ways that capture everything mathematicians care about (if not everything philosophers care about).

Third, I think there's some awkwardness in the fact that Shapiro's articulation of Structuralism would seem to allow that studying the weird metaphysical properties of his special objects called structures would count as mathematics.

In contrast, I've suggested a different way of formulating Shapiro's Structuralist thesis: that mathematical questions can all be formulated as questions pure logical possibility.[7] This yields what I take to be a more intuitive verdict: that studying the metaphysical properties of either Shapiro's structures qua exotic abstract objects or my primitive modal notion of conditional logical possibility (e.g., is it a reference magnet? does taking conditional logical possibility to be a metaphysical primitive commit us to Tractarian/Russellian ideas about there being a preferred carving of the world?) does not count as mathematics. For, these claims about the nature of logical possibility cannot themselves be formulated as pure logical possibility claims (i.e., sentences of the form $\Box\Phi$ or $\Diamond\Phi$ where the $\Box$ and $\Diamond$ operators are unsubscripted and Φ is a sentence in the language of logical possibility).

---

[7] See Chapter 10.

18 Anti-Objectivism About Set Theory

So far, we have discussed Actualist and Potentialist approaches to set theory. In this chapter I will discuss a third major family of approaches to set theory: anti-Objectivist views on which some questions in the language of set theory lack determinate right answers. Such views are fairly popular. For example, many philosophers and mathematicians find it plausible that there's no fact of the matter about the Continuum Hypothesis (the claim that there is no set whose cardinality is strictly between that of the integers and the real numbers), which has famously been shown not to be provable or refutable from the ZFC axioms of set theory.

In this chapter, I will discuss some major examples of anti-Objectivist philosophies of mathematics, and some concerns that arise for them. After reviewing a useful distinction between strong and weak anti-Objectivism about set theory, I'll discuss Field's remarks about weak anti-Objectivism, Feferman's Social Constructivism (a form of weak anti-Objectivism) and Hamkins' Multiverse Program (possibly a form of strong anti-Objectivism). In doing this, I hope to offer some support (or at least explanation) for my choice to handle arbitrariness worries in Chapter 2 by embracing Potentialist set theory rather than some anti-Objectivist option.

18.1 Strong vs. Weak Anti-Objectivism

We can distinguish two different types of anti-Objectivism about set theory:

- Strong anti-Objectivist approaches to set theory hold that all undecidable sentences in set theory (i.e., sentences which can't be proved or refuted using our best total mathematical theory) are neither determinately true nor determinately false.
- Weak anti-Objectivist approaches to set theory hold that *typical* undecidable sentences in set theory are neither determinately true nor determinately false.

So, for example, if CH qualifies as a "typical" undecidable sentence, then both types of anti-Objectivist theories will say neither CH nor ¬CH will be determinately true.

As Field (1998) usefully notes, being an anti-Objectivist doesn't prevent you from using classical logic. For example, even strong anti-Objectivists can still say that CH ∨ ¬CH is determinately true, since this disjunction *is* provable in our best mathematical theories.

Also, anti-Objectivism about set theory can be combined with either realism (i.e., Platonism) or anti-realism about mathematical objects. For example, a philosopher who combines Platonism and anti-Objectivism might say there's a large mathematical universe containing many different hierarchy-of-sets like structures. All of these structures satisfy all our principles of first order set theory. However, for any undecidable sentence of set theory ϕ, there will be a hierarchy of sets V_1 in this mathematical universe which makes ϕ true and another hierarchy of sets V_2 which makes it false. A set-theoretic sentence will be determinately true if and only if it is true on all of these models (which all provide equally legitimate interpretations of our set-theoretic concepts). Thus, undecidable sentences won't be determinately true or false. However, sentences of the form $\phi \vee \neg\phi$ will be determinately true, since it is true in all the relevant structures.[1]

Some versions of anti-Objectivism (at least) promise to let us avoid the arbitrariness worries discussed in Chapter 2.[2] For the Nominalist Anti-Objectivist can deny that there is a single hierarchy of sets (or a plurality of hierarchies of sets) which happens to stop at some point which is not determined by our conception of the hierarchy of sets (and thus qualify as rivals to the Potentialist set theory advocated in this book).

18.1.1   Problems for Strong Anti-Objectivism

Field notes that we can use Putnamian model theoretic reference skepticism to motivate strong anti-Objectivism. Crudely speaking, one might argue that since we lack causal contact with mathematical objects, the only thing which can constrain the reference of our terms like "set" is our best first order mathematical theory about the sets. Thus, there's a prima facie question of how any sentence ϕ such that both ϕ and $\neg\phi$ are compatible with our best first-order theory of the sets could be determinately true or false.

However, Field then plausibly argues that strong-anti-Objectivism faces problems about what to say concerning mathematical consistency facts, as follows.

There's compelling reason to believe that our best total first order theory of set theory is recursively enumerable. Thus, by Gödel's famous theorem (Gödel 1931) there will be some number theoretic sentences that aren't decidable by our best total theory. So, the strong anti-Objectivist will say that some number theoretic sentences aren't determinately true or determinately false.

---

[1] In comparison, a Nominalist anti-Objectivist might say that (if T is an axiomatization of all our first-order beliefs about set theory) an arbitrary set-theoretic sentence ϕ is:

- determinately true iff ϕ is provable (using standard first-order logic) from T;
- is determinately false iff $\neg\phi$ is provable from T;

and otherwise, indeterminate.

[2] It's not clear that Platonist anti-Objectivist theories help with arbitrariness. For although they're not committed to any single height, they do seem to be committed to a stopping point in a different sense. It is appealing to think that for any plurality of objects, it would be logically possible for there to be a hierarchy of sets like structure which contains a set of this plurality (and thus adds to it). In this case, the Platonic anti-Objectivist is committed to positing an arbitrary stopping point (a logically possible upper bound past which none of our hierarchy of sets goes) just as much as the standard Platonist set theorist is.

But this can seem unintuitive. For one thing, many people have the intuition that there should be determinate right answers to *all* questions of number theory. More worryingly however, even if we are willing to bite this bullet, saying that there aren't facts about what Con sentences (i.e., number theoretic statements corresponding the claim that no number Gödel codes a proof of "0=1" from some certain mathematical axioms) are true has powerful effects on how we can understand the rest of the strong view.

The issue is that saying some con sentences are indeterminate creates pressure to say there aren't always determinate facts about what sentences are provable (i.e., provable by a finite series of applications of the first order logical inferences to our best theory). For suppose there are determinate facts about provability, i.e., for every first-order theory T either "$0 = 1$ is provable from T" is determinately true or it is determinately false. But $\text{Con}(T)$ is simply a formalization of the claim that it's not the case that "$0 = 1$ is provable from T"[3] and most people treat the biconditional claim connecting provability and the existence of a natural number coding for a proof as clearly true (and perhaps even something like analytic).[4] So, one might think that these biconditionals must come out true on all acceptable interpretations of our mathematical talk. Thus, from the assumption that all claims about provability are determinately true or false it plausibly follows that the corresponding con sentences must be determinately true or false (i.e., it's true if $0 = 1$ is provable and false if not).

But then saying that there aren't determinate facts about what's provable undermines our grip on the stated definition of strong anti-Objectivism: the claim that set-theoretic sentences are determinately true iff they are provable from our best physical theory T. Are we now to say that, not only is there no determinate fact about certain mathematical claims, but there isn't even a determinate fact about what claims have a determinate right answer (i.e., are provable)?

Field notes that we can fix this problem by replacing strong anti-Objectivism with an even more extreme anti-Objectivism, which says that only sentences provable in less than n stages (for some n larger than any number of stages a proof could actually contain) are determinately true. But, of course, this is a very radical move and any particular choice of n looks unmotivated.

18.2　Weak Anti-Objectivism

Weak anti-Objectivism lets us avoid the problem described in the last section. For the weak anti-Objectivist can say there are determinate facts about provability and right answers regarding mathematical Con sentences, but more typical/complex set-theoretic questions that can't be settled by proof lack right answers.

---

[3]　That is, $\text{Con}(T) \overset{\text{def}}{\leftrightarrow} \neg(\exists n)\text{Proves}(0 = 1, T, n)$ where $\text{Proves}\,(0 = 1, T, n)$ asserts that n doesn't code a proof from T. Though, formally speaking, the predicate would accept a code for a computable axiomatization of T in those cases T is infinite like the ZFC axioms.

[4]　If pressed to explicate when we should say that a particular statement is provable we would cash that out in something like the existence of a finite number of steps, each of which follows from the next according to some finite list of rules, and we'd agree that such a sequence exists just if there is an integer appropriately coding it.

The weak anti-Objectivist can even keep the Putnamian model theoretic motivations invoked in the last section, if they can find a way to say that something (e.g., physical reality) pins down a unique interpretation for talk of the natural numbers (and hence determinate right answers to all Con sentences), but there are still different equally good interpretations for the rest of set theory, so typical set-theoretic claims not decidable from our axioms don't have definite right answers.

Field (1998) cautiously proposes a specific way of implementing these ideas – pinning down determinate reference to a natural number structure (up to isomorphism). In a nutshell, he is using determinate physical reference and facts about space to pin down a definite notion of finitude (which lets us rule out nonstandard models of PA, pin down a determinate truth-value for all number theoretic sentences and hence all claims about provability).[5]

The resulting view about set theory, which I will provisionally call Field's weak anti-Objectivism, has significant attractions. For example, it provides clear answers to certain Putnamian questions about how we can refer to mathematical structures. Additionally, Nominalist implementations of this anti-Objectivism promise to avoid the arbitrariness worries for Actualism discussed in Chapter 2, while still vindicating set theorists' use of classical logic.

18.2.1 Objections to Field's Weak Anti-Objectivism

I will now explain why I take my Potentialist approach to be preferable.

My main objection to Field's proposal concerns his Putnamian motivation for anti-Objectivism. It is beyond the scope of this book to try to refute Putnamian reference skepticism. However, it should be noted that Putnam's arguments for reference skepticism are very general, also blocking determinate reference to physical objects and suggesting that (in a certain sense) we can't be undetectably wrong about the sciences. So those who accept scientific realism are already committed to something's being wrong with Putnam's argument.[6]

Second, one might note that, in *Saving Truth from Paradox*, Field (2008a) argues that we should accept a notion of logical possibility as a conceptually primitive and seems to allow determinate reference to it. But if we can somehow secure determinate reference to logical possibility simpliciter, it seems natural that we should also have determinate reference to the notion of conditional logical possibility which generalizes it. And we have seen the latter to suffices to pin down a unique intended width to the

---

[5] I raise some issues with this and propose my own argument for a relevant conditional claim in Berry (2020b): if we can somehow secure determinate realist reference to physical notions plus notions of physical possibility, we can pin down a unique natural number structure. Though, I only argue for the conditional claim, not that we can secure such definite reference.

[6] One might argue that causal contact with some physical objects can pin down determinate reference to them in a way that it could not pin down reference to mathematical objects or a preferred notion of logical possibility. But if you accept that we can refer to (and ideal science can be wrong about) determinate non-Humean facts about physical possibility or objective physical probability, this line is difficult to take. For it doesn't seem like causal contact can play a very different role in explaining determinate reference to these modal notions than it can in explaining determinate reference to logical possibility.

hierarchy of sets – and determinate right answers to all set-theoretic claims in a Potentialist framework.

18.3   Feferman's Conceptual Structuralism

Next let us turn to Feferman. In works like Feferman (2011) he advocates a view called Conceptual Structuralism. On this view mathematics studies mental conceptions. These are socially constructed objects, like marriages, bank balances and contracts. And their properties can change as mathematicians refine their way of thinking about, e.g., the real numbers. He writes:

The basic objects of mathematical thought exist only as mental conceptions, though the source of these conceptions lies in everyday experience in manifold ways, in the processes of counting, ordering, matching, combining, separating, and locating in space and time. (Feferman, 2011)

Mental conceptions are world pictures which describe structures, that is, "coherently conceived groups of objects interconnected by a few simple relations and operations" (Feferman, 2011) and exist prior to any choice of axioms or logical development.

According to Feferman, the aim of mathematics is to start with some features of these structures which our conceptions make obvious and then work out further features. Importantly, conceptions of mathematical structures can be more or less clear, and when they are not fully clear there can be failures of determinate truth-value.

In particular, Feferman suggests that our conception of the natural numbers via thinking about stroke sequences is fully clear. On the other hand, our conception of the real numbers (particularly our set-theoretic conception of the real numbers) is not fully clear. And indeed, it cannot be made clear without violating the kinds of plenitude intuitions which belong to the conception right now (e.g., we can't just stipulate that all subsets of the natural numbers are definable/constructable in certain ways). Thus, there is no fact of the matter about questions like the Continuum Hypothesis, because there are no determinate facts about how many sets something containing "all possible subsets" of the natural numbers would have to have.

I take Feferman's proposal to be a form of weak anti-Realism. For it allows there to be definite right answers to all sentences in the language of arithmetic. But it denies that there are definite right answers to some other questions stated in the language of set theory.

Accepting Feferman's social constructive approach to set theory would let us avoid the arbitrariness problems for Actualism noted in Chapter 2. For Feferman could (and would) say that our conception of the hierarchy of sets is not fully clear with regard to the intended height of the sets. And, unlike the Platonist, he does not take there to be Platonic objects forming a hierarchy of sets which extend up a certain height but no further.

So why do I favor my Potentialist approach to set theory over Feferman's? First, Feferman's main cited reason for preferring his view to Platonism is the Benacerraf problem. He writes, "The assumption of all these definite totalities [(namely the powerset of the natural numbers and the powerset of that)] is only justified by Platonic realism" and then notes that realism faces the Benacerraf problem. So, it won't apply if you're optimistic about the account of possible human access to logical possibility facts discussed in Chapter 19 and Berry (2018b).

Second, as Peter Koellner (2016) crisply puts it, Feferman's conception of "adequate clarity" can itself seem troublingly unclear. Feferman seems interested in the possibility of forming various mental pictures, like the stroke sequence associated with our conception of the natural numbers. And he writes that our conception of the continuum in terms of a line is clearer than the set-theoretic conception of it in terms of arbitrary subsets of the integers. Feferman writes:

we have a much clearer conception of arbitrary sequences of points on the Hilbert (or Dedekind, or Cauchy-Cantor) line, or at least of bounded strictly monotone sequence, than we do of arbitrary subsets of the line. And ... we have a clearer conception of what it means to be an arbitrary infinite path through the full binary tree than of what it means to be an arbitrary subset of N, but in neither case do we have a clear conception of the totality of such paths, resp. sets. (Feferman, 2011)

But having a clear conception doesn't just mean being able to have some mental picture of that structure, e.g., imagining the hierarchy of sets by mentally picturing a V-shaped expanding column. Feferman suggests that there's a way in which this picture represents itself as being fully determinate, yet fails to be so determinate:

There is no problem to put oneself in the mental frame of mind of "this is what the cumulative hierarchy looks like," for which one can see that such and such propositions including the axioms of ZFC are (more or less) obviously true. I have taught set theory many times and have presented it in terms of this ideal-world picture with only the caveat that this is what things are supposed to be like in that world, rather than to assert that's the way the world actually is. (Feferman, 2011)

Additionally, note that Feferman's notion of having a fully definite conception of a mathematical structure cannot be identified with being able to prove or refute all statements about that mathematical structure. And, more generally, he can't mean that a conception is definite about whether ϕ iff we can prove or refute ϕ by contemplating our conception of the structure. For Feferman says that our conception of the natural numbers is fully determinate, and by standard points about the incompleteness theorem (and the plausibly computable nature of the human mind) there are questions about the numbers we won't be able to decide by eliciting them from this conception.

Third, I wonder whether Feferman's view can make sense of certain applications of mathematics. He can make sense of the applications of number theory to what physical structures exist, by saying that these instantiate the number structure we can conceive by imagining strokes. But suppose that we had such a physical structure. And suppose that our conception of subsets of the natural numbers structure left it indefinite whether there is any subset of the numbers with some property ϕ. Now suppose some physical property, like being purple, applied to the objects in this stroke sequence in such a way that the physical strokes which were purple had this property. In this case, intuitively, the claim that there is a subset of the numbers that satisfies ϕ would have to be determinately true, despite not being adequately pinned down by our conception.

18.4   Hamkins

Set theorist Joel Hamkins has developed an influential[7] multiverse approach to set theory, on which there are many different hierarchies of sets, and there's no fact of the matter about whether certain set-theoretic statements are true, beyond the fact that they are true of some hierarchies of sets within the multiverse and false in others. On this view there is no full intended hierarchy of sets which contains all subsets of sets it contains – or even all subsets of the natural numbers. Rather, for *every* set-theoretic universe V in the multiverse, there's an extending "fatter" hierarchy of sets $V[G]$ that includes all sets in V but also an extra "missing subset" of the set of natural numbers in V.

I will only have time to scratch the surface of responding to this program here.[8] I will discuss the motivations for the Multiverse view that Hamkins (2013) provides. And I'll suggest that despite Hamkins central use of an analogy between set theory and geometry to motivate his set-theoretic multiverse program, there are important limits to this analogy which raise a (prima facie) explanatory indispensability worry for his view.

18.4.1   Hamkins' Multiverse

18.4.1.1   Forcing Fundamentals

Before describing Hamkins' multiverse, I will first review some very basic mathematical facts about forcing, as this technique plays a central role in Hamkins' program (and some of these mathematical facts will play an important role in my argument). In particular, Hamkins' main motivation for the multiverse, aside from the analogy with geometry, arises from the idea that we should understand mathematical arguments by forcing in a certain unconventional way.

Set-theoretic forcing was, famously, developed by Paul Cohen to prove the independence of the Continuum Hypothesis (i.e., the claim that there is no set intermediate in size between the real numbers and the natural numbers). However, this method has been generalized to prove a broad range of meta-mathematical results.

As standardly presented, forcing is a technique which lets one produce a new model of set theory from an original *countable* well-founded[9] model M of set theory.

We work in the total hierarchy of sets V, which we assume to satisfy the ZFC axioms of set theory. And we consider an infinite partial order $\mathbb{P}$ that is a set in our countable model M. Because M is countable, it has to be missing some subsets of any infinite set $\mathbb{P}$ it contains, by Cantor's diagonal argument. Our strategy will be to expand M by adding a missing subset of this set $\mathbb{P}$ to M.

---

[7] See, for example, Gaifman (2012); Button and Walsh (2016); Pruss (2019); Clarke-Doane (2020); Jonas (2020); Koellner (n.d.).

[8] Many thanks to Peter Gerdes for help with this section, and thanks to Peter Koellner for much relevant lecture and informal conversation.

[9] More specifically, forcing lets you produce a new model of set theory extending every countable *transitive* model of set theory. A model M of set theory is transitive iff the membership relation in M is $\in$, i.e., $x \in_M y \leftrightarrow x \in y$. However, by the Mostowski collapse lemma (Jech 1981), any well-founded countable model is isomorphic to a transitive model.

Specifically, we can use the fact that M is countable to prove that there's an "M-generic" set $G \subset \mathbb{P}$ (where being M-generic implies *not* being a set in M).[10]

Next, we consider a fatter model of set theory $M[G]$ which expands M by adding G to it (along with other sets, as needed to satisfy the ZFC axioms.[11] Finally, we show that any such $M[G]$ must satisfy some desired claim ϕ. In this way we prove the relative consistency of ZFC $+ \phi$.

But now the key point about forcing arguments that opens the door to Hamkins' multiverse is this. The mechanics of forcing allow us to make claims that *only quantify over* sets in original countable model of set theory[12] M but can be seen as *implicitly telling us about* this larger model of set theory $M[G]$[13] in the following sense.

We can define a relation $\Vdash$ (called a **forcing relation**) such that the claim that $\Vdash \phi$ only involves sets in M but we can prove the following biconditional (without appeal to the fact that M is countable). If there is any M-generic subset of $\mathbb{P}$:

$\Vdash \phi$ if and only if $M[G] \models \phi$ for every generic $G \subset \mathbb{P}$.

That is, a sentence ϕ is forced ($\Vdash \phi$) if and only if for any M generic subset G of P, ϕ is true in the expanded model of ZFC $M[G]$ we get by adding G.

A specific forcing argument proceeds by picking an infinite partial order $\mathbb{P}$, which we will add a subset of, and then proving that $\Vdash \phi$ holds when ϕ is some claim we wish to show is consistent with the ZFC axioms.

So, for instance, Cohen proved in ZFC that there is a partial order $\mathbb{P}$ such that $\Vdash \neg CH$ (where CH is the Continuum Hypothesis). Thus, if M is a *countable* transitive model of ZFC, then (if G is a generic object for $\mathbb{P}$), $M[G]$ is a countable transitive model of ZFC $+ \neg$ CH. Of course, speaking formally, we can't assume that there are *any* models of ZFC, but this is enough to establish the consistency of ZFC$+\neg$ CH provided we think ZFC is consistent (and hence has a countable model).

18.4.1.2   Hamkins' Proposal

Hamkins describes his multiverse proposal as a form of Platonism:

> The multiverse view is one of higher-order realism—Platonism about universes—and I defend it as a realist position asserting actual existence of the alternative set-theoretic universes into which our mathematical tools have allowed us to glimpse. (Hamkins 2012)

However, rather than accepting a single platonic hierarchy of sets, he proposes that there are many different hierarchies of sets. The set-theoretic multiverse is the space of all such set-theoretic hierarchies. And certain set-theoretic statements, like the

---

[10] Specific, a generic, i.e., a generic filter G is a filter which intersects every dense subset of $\mathbb{P}$ included in M.

[11] $M[G]$ winds up being the smallest transitive model of ZFC extending M and containing G as a set.

[12] That is, model of ZFC.

[13] While M can't define truth in $M[G]$ in M one can define a class of names for objects in $M[G]$ (some of which may refer to the same object) and a forcing relation $p \Vdash \phi$ (where ϕ is a sentence in the language of set theory and p an element of the forcing partial order $\mathbb{P}$ supplemented with the aforementioned class of names), which holds just if $M[G] \models \phi$ for every generic object G containing p.

Continuum Hypothesis are not true or false simpliciter, but merely true in some parts of the multiverse and false in others.

As Hamkins vividly argues in the passage below, CH cannot be settled by finding intuitively compelling new axioms from which it can be proved or refuted, because mathematicians' experience reveals there are parts of the multiverse in which *CH* holds and parts in which ¬CH holds:

> [If some obviously true seeming mathematical axiom] ϕ were proved to imply CH, then we would not accept it as obviously true, since this would negate our experiences in the worlds having ¬CH. The situation would be like having a purported "obviously true" principle that implied that midtown Manhattan doesn't exist. But I know it exists; I live there. Please come visit! Similarly, both the CH and ¬CH worlds in which we have lived and worked seem perfectly legitimate and fully set-theoretic to us, and because of this, any proof from ϕ that CH or that ¬CH casts doubt to us on the naturality of ϕ. (Hamkins 2012)

Hamkins' view of the multiverse is heavily influenced by the set-theoretic technique of forcing just described. In particular, he suggests that for any set-theoretic hierarchy V we should accept that (for an appropriate partial order $\mathbb{P}$ in V) there is another set-theoretic hierarchy $V[G]$ corresponding to the forcing extension of V with respect to the partial order $\mathbb{P}$. As we saw in Section 18.4.1.1, this claim is straightforwardly true if we work in some background notion of set theory and take V to be a *countable* model of set theory. But Hamkins suggests that we should assume that any (forcing appropriate) partial order admits a generic filter, so that even whatever total set-theoretic hierarchy V we are currently working in has a forcing extension. He writes:

> [A] set theorist with the universe view can insist on an absolute background universe V, regarding all forcing extensions and other models as curious complex simulations within it. (I have personally witnessed the necessary contortions for class forcing.) Such a perspective may be entirely self-consistent, and I am not arguing that the universe view is incoherent, but rather, my point is that if one regards all outer models of the universe as merely simulated inside it via complex formalisms, one may miss out on insights that could arise from the simpler philosophical attitude taking them as fully real. (Hamkins 2012)

This claim that we can extend *every* set-theoretic structure by taking a forcing extension is a crucial and deeply controversial aspect of Hamkins view. For it directly conflicts with the standard realist intuition that it's possible to build a set-theoretic hierarchy that already contains "all possible subsets" of any set in that hierarchy (c.f. IHW in Section 2.2). For any such set-theoretic hierarchy V must already contain all subsets of every partial order $\mathbb{P}$ it contains. Thus, there should not be any generic $G \subset \mathbb{P}$ which isn't a member of V, i.e., V and $V[G]$ should always be the same. For instance, if one thinks that a set-theoretic hierarchy already contains all possible subsets of the integers, it would be impossible to extend that hierarchy via a forcing extension which adds another subset of the integers.

While Hamkins' proposal seems to take significant motivation from the example of forcing extensions, this isn't the only closure principle about the multiverse which he accepts. It isn't even the most controversial. He also suggests that every set-theoretic hierarchy V is countable from the perspective of some other hierarchy V'

(Hamkins 2012). Indeed, he suggests that – although "this principle appears to be abhorrent to most set theorists" – every set-theoretic hierarchy V is ill-founded from the perspective of another set-theoretic hierarchy V'.

18.4.2   Motivating the Multiverse

Why should one accept this radical approach to set theory? In this section, I'll discuss two motivations Hamkins gives in his philosophical overview "The Multiverse Perspective in Set Theory" (Hamkins 2013), then raise some worries for his approach.

18.4.2.1   Mathematical Practice and Phenomenology

First, Hamkins appeals to the practice and phenomenology of set theory (in particular forcing arguments). He notes that now, rather than stating results proved by forcing as consistency claims of the form $\text{Con}(\text{ZFC} + \phi) \to \text{Con}(\text{ZFC} + \psi)$, "contemporary work would state the theorem as: if ϕ, then there is a forcing extension that satisfies ψ." The latter claim could be read as asserting the existence of a forcing extension *of the total set-theoretic hierarchy V you are working in*, rather than any mere countable model M satisfying $\text{ZFC} + \phi$. Hamkins' Multiverse hypothesis takes this appearance at face value.

Hamkins also appeals to the phenomenology of making forcing arguments, which he describes as follows (emphasis mine), and suggests that his multiverse proposal takes at face value:

> [The multiverse proposal] makes sense of our experience—in a way that the universe view does not—simply by filling in the gaps, by positing as a philosophical claim the actual existence of the generic objects which forcing comes so close to grasping, without actually grasping. With forcing, we seem to have discovered the existence of other mathematical universes, **outside our own universe,** and the multiverse view asserts that yes, indeed, this is the case. We have access to these extensions via names and the forcing relation, even though this access is imperfect. Like Galileo, peering through his telescope at the moons of Jupiter and inferring the existence of other worlds, catching a glimpse of what it would be like to live on them, set theorists have seen via forcing that divergent concepts of set lead to **new set-theoretic worlds, extending our previous universe,** and many are now busy studying what it would be like to live in them. (Hamkins 2013: 11)

Equally eminent set theorists who reject the multiverse program (Martin 2001) might give a different description of the phenomenology. And even if one grants this point, it's disputable whether the multiverse better fits mathematical practice and phenomenology than conventional realist approach to set theory (paired with the conventional interpretation of forcing) *overall*.

Admittedly Hamkins' proposal accords better with intuitions that forcing arguments reveal, "new set-theoretic worlds, extending our previous universe." However, one might argue that traditional realist approaches to set theory account for many more aspects of mathematical intuition and practice overall than Hamkins' theory does. For,

Hamkins admits that his own principles about what hierarchies exist in the multiverse will be "abhorrent to many set theorists."

18.4.2.2 An Analogy Between Set Theory and Geometry

However, perhaps a second kind of motivation which more clearly supports the multiverse perspective over more traditional realism can be found later in Hamkins (2012). For Hamkins dramatizes the kind of view that he takes set-theoretic practice to support (and the subsequent change in mathematical attitudes he implicitly advocates) by drawing an analogy between set theory and geometry. If the multiverse view could be shown to fall out of treating set theory and geometry similarly, this could powerfully motivate accepting it.

I will quote Hamkins' specific and somewhat unusual development of this familiar comparison at some length because of its importance to the argument below. He writes:

There is a very strong analogy between the multiverse view in set theory and the most commonly held views about the nature of geometry. For two thousand years, mathematicians studied geometry, proving theorems about and making constructions in what seemed to be the unique background geometrical universe. In the late nineteenth century, however, geometers were shocked to discover non-Euclidean geometries. At first, these alternative geometries were presented merely as simulations within Euclidean geometry, as a kind of playful or temporary reinterpretation of the basic geometric concepts. For example, by temporarily regarding "line" to mean a great circle on the unit sphere, one arrives at spherical geometry, where all lines intersect; by next regarding "line" to mean a circle perpendicular to the unit circle, one arrives at one of the hyperbolic geometries, where there are many parallels to a given line through a given point. At first, these alternative geometries were considered as curiosities, useful perhaps for independence results, for with them one can prove that the parallel postulate is not provable from the other axioms. In time, however, geometers gained experience in the alternative geometries, developing intuitions about what it is like to live in them, and gradually they accepted the alternatives as geometrically meaningful. Today, geometers have a deep understanding of the alternative geometries, which are regarded as fully real and geometrical. (Hamkins 2013)

In this quote, Hamkins compares set theorists who approach forcing conventionally (as studying countable models inside the true intended hierarchy of sets V) to old geometers who took studying non-Euclidean geometries to be legitimate mathematics but only to reveal syntactic facts about provability and consistency, plus what would be true under some "playful reinterpretations" of the terms "point" and "line" in these axioms.

Hamkins suggests that set theorists should mirror geometers' eventual adoption of alternate axiom systems as "geometrically meaningful" and "alternate geometries ... as fully real" and that adopting the Multiverse theory corresponds to doing this. Rather than seeing alternative set-theoretic axiom systems as having mere countable toy models within a larger system, we should (as per the change in attitude to forcing Hamkins advocates) see alternative axiom systems as describing something more fully real and genuinely set-theoretic by seeing them as genuine extensions of whatever hierarchy of sets we are currently working in.

18.4.3   Questions and Concerns

With this overview of Hamkins' multiverse and his motivations for it in mind, I want to raise a worry.

In the case of geometry, in addition to the questions about various geometries which can be studied in pure mathematics (as Hamkins mentions), one might say there's a further question: what's the geometry of the physical space we live in?[14] The change of opinions about geometry alluded to in Section 18.4.2 didn't deny the existence of robust metaphysically joint-carving laws with essentially the same physical consequences naive geometry had claimed. It just downgraded these laws from metaphysical necessities to mere physical necessities. Appeal to physical geometry provides an important sense in which, e.g., the parallel postulate is definitely false (which is not relative to a choice of axioms to work in). In contrast, Hamkins doesn't seem to allow that there's a physically preferred set theory – a set theory that takes over the traditional a priori applications of naive set theory.

So, in the case of geometry, we have two things: a range of different geometries which constitute equally legitimate objects for (non-formalist) mathematical study *and* (fairly) determinate facts about which geometry corresponds to our physical reality. Hamkins' multiverse proposal nicely mirrors the former idea, but he does not seem to accept anything corresponding to the latter.

This raises a potential explanatory indispensability problem. In the case of geometry, we say there's an important joint-carving notion (physical geometry), that can take over the scientific-explanatory work done by appeals to naive geometry (explaining physical facts by appeal to *some kind* of genuine counterfactual-supporting laws) and explain the attraction of naive geometry. But what can Hamkins say about the analogous phenomenon of apparent scientific-explanatory appeals to set theory? That is, what can he say about scientific explanations that seemingly appeal to a preferred notion of "all possible ways of choosing" from a given plurality of objects/conditional logical possibility facts to explain physical regularities (like the three-colorability example and others discussed in Chapter 13)?

When explaining why the infinite map considered in my toy explanation isn't three colored, the naive set theorist appeals to a hierarchy of sets with ur-elements, and the idea that it contains sets corresponding to "all possible ways of choosing" elements of the sets it contains. Since there is a set of all ordered pairs of countries and colors, if there were a three coloring, there would have to be a set witnessing it. But in Hamkins' multiverse there can be no such set-theoretic hierarchy (each hierarchy sits within a larger one that adds extra subsets). The appearance that there's a distinguished notion of all possible subsets/all possible ways of choosing some colors on the map is an illusion.

Thus, Hamkins faces a kind of prima facie (not to say insolvable) explanatory indispensability problem. And, interestingly, this problem is raised by his mathematical

---

[14] That is, there are facts constraining the behavior of all actual spatial points and lines etc., as well as facts about what's possible within various alternate geometries we can metaphorically visit and imagine living in by doing mathematics with different axioms.

anti-Objectivism/truth-value anti-realism (his rejection of an intrinsically preferred notion of all possible subsets) rather than the rejection of mathematical objects which traditional mathematical indispensability arguments target.

Additionally, even if Hamkins can answer the above explanatory indispensability problem, I take this point about the a priori intended physical applications of current ("naïve") set theory to show that Hamkins is advocating more radical change in attitudes to set theory than the change in attitudes to geometry he invokes to motivate it. He can't say that the multiverse proposal is just what falls out when we treat set theory the way that we've learned to treat geometry. In the geometrical case we allowed that there was a legitimate notion of spatial possibility and there are genuine important (perhaps even very precisely determinate) facts about what's allowed by the geometry of space. The useful explanatory distinctions and counterfactual-supporting laws invoked by naive geometry just turned out to be out more parochial and less a priori than had been thought[15] and no longer seemed to deserve the unique mathematical status they'd previously been accorded. In contrast, Hamkins' multiverse program seems to suggest that the appearance that there's a joint-carving distinction in the neighborhood of our naive talk about "all possible ways of choosing" (and thence the intended model of set theory, up to width) is an illusion.

Additionally, why does Hamkins take understanding forcing as telling us about $V[G]$ to correspond to regarding variant axiom systems as describing something "fully real" and "set-theoretic" in a way that treating forcing as merely telling us about some countable model $M[G]$ would not? After all, both structures are made of sets that literally exist, and satisfy the ZFC axioms.

---

[15] That is, it turned out that we were appealing to physically necessary laws of space not mathematically necessary ones.

19 Conclusion

19.1 Summary

The idea that there's some intimate connection between mathematics and logic has been a central focus of attention within analytic philosophy since Frege. In this book I have developed a specific version of this idea, suggesting we can solve certain philosophical problems by thinking about set-theoretic statements in Potentialist terms, i.e., as abbreviating certain modal claims about logical possibility and extensibility.

In Part I, we saw that Actualist approaches to set theory faced an arbitrariness problem arising from the Burali-Forti paradox, concerning our conception of the height of the hierarchy of sets. Potentialist approaches to set theory would let us avoid this problem, but raise their own worries concerning, e.g., how the notion of possibility which is used to paraphrase claims of set theory should be used. Both Actualist and Potentialist theories faced a problem about how to justify acceptance of the ZFC axioms, particularly the axiom of Replacement.

I tried to point out an intuitive notion of conditional logical possibility and argued that it could be attractively taken as a conceptual primitive. I formulated a version of Potentialist set theory which employed this modal notion and (unlike existing versions of Potentialism) avoided quantifying-in to the $\Diamond$ of logical possibility. I then argued that this formulation simplified existing formulations of Potentialist set theory and solved some of the philosophical problems for them discussed in this book (e.g., see Chapter 2).

In Part II, I proposed a formal inference system for reasoning about conditional logical possibility. I then showed this interference system can be used to justify Potentialist acceptance of all theorems in ZFC on the basis of general methods of reasoning about logical possibility which feel as intuitively obvious (once one understands them), as we would naively hope the building blocks of mathematical arguments to be.

In this way, I have suggested that we could usefully sharpen our thinking about pure set theory (in contexts where potential philosophical confusions matter) by thinking about set-theoretic expressions as abbreviating Potentialist claims. I have also argued that if we do so, it is harmless for mathematicians to simply reason from ZFC using first-order logic (without considering or expanding out Potentialist logical analyses of these claims).

In Part III, I discussed how the understanding of set theory defended in the bulk of this book might be extended to a more general philosophy of mathematics. En route, I tried to highlight some interesting heterogeneity in the role of mathematics in the sciences and show how philosophical analyses using the logical possibility operator might be useful to philosophy of language, metametaphysics and general epistemology.

In this chapter I will conclude by making some very brief suggestions about what larger consequences all this could have.

19.2 Truth-Value Realism

The first consequence concerns commonplace questions about mathematical truth-value realism, e.g., are there definite right answers to mathematical questions which have been shown to be independent of the ZFC axioms of set theory (or questions we could never settle by proof)?

It is controversial whether statements like the Continuum Hypothesis and the generalized Continuum Hypothesis[1] (which have been shown to be independent of the ZFC axioms (Cohen 1963)) have determinate truth-values. By reducing these statements to claims stated purely in terms of first-order logic and logical possibility I provide, at least for those who accept there are definite facts about logical possibility, reason to accept these statements have a definite truth-value.

Accepting the version of Potentialist set theory I have advocated in this book motivates a blanket positive answer to all questions about whether statements in the language of first-order set theory have determinate truth-values. For my Potentialist translation transforms these mathematical puzzles into pure questions in the language of logical possibility. And it is appealing to think that we can refer to a unique intended notion of logical possibility such that there are definite right answers to these questions of pure logical possibility.

Of course, Hellman and Linnebo also provide a reduction of these set-theoretic claims to statements involving their preferred modal notions. Similarly, if one takes Potentialist set theory to appeal to a *sui generis* notion of constructability or interpretational possibility (as Linnebo does) there is (arguably) less reason to assume that facts about this modal notion will behave classically. Also, insofar as Hellman's or Linnebo's Potentialist paraphrases use quantifying-in, one might accept a Potentialist understanding of set theory but take failures of determinacy to trace back to these. That is, if you doubt (as some do – see Quine 1953) that there are clear truth-conditions for quantifying-in you would similarly doubt that the Potentialist translations given by Hellman and Linnebo suffice to give truth-values to these claims.

---

[1] The Continuum Hypothesis says that nothing can have a size between that of the natural numbers and the real numbers. The Generalized Continuum Hypothesis says that for each cardinality α there is no possible cardinality β between α and 2^α.

19.3   The Access Problem

The second larger consequence I'd like to draw attention to concerns a classic challenge to mathematical realism which is sometimes called the access problem or the Benacerraf problem (Benacerraf 1965). In general, the access problem asks how, if realist philosophies of mathematics are correct, human accuracy about mathematics can be anything but a miracle or a mystery (given that, e.g., we cannot see or touch or taste or otherwise causally interact with mathematical objects).

In the specific case of set theory, we can ask: if set theorists are really getting at objective (but abstract and causally inert) proof transcendent facts in the way that you, the realist, say that it is, how can this match between objective reality and human psychology be anything but a miracle or a mystery? If one accepts my interpretation of set theory, then this challenge becomes a challenge to understand how human beings can have accurate methods of reasoning about logical possibility and extensibility.

And if one accepts the story about mathematical objects other than the sets developed in Part III, then access worries about all of mathematical knowledge can be reduced to access worries about knowledge of logical possibility.

Note that, as I have emphasized at various points in this book, the laws of logical possibility are supposed to be "subject matter neutral," constraining the behavior of all objects and relations – from numbers to apples to ghosts or genres of novels. So, there's some hope that we could (in effect) abductively learn general laws of logical possibility from dealing with non-mathematical objects, and then apply them to deduce possibility claims about the very large and complex structures studied in (Actualist or Potentialist) set theory – much as one could learn laws of physical possibility from experiments on the earth with pendulums etc. and then apply them in space.

Thus, overall, one might hope the access problem for (pure) facts in the language of logical possibility is relatively tractable, and continuous with analogous access worries/questions we might ask about human knowledge of physical, psychological or chemical possibility. However, telling that story is a project for another book.

Appendices

An online appendix with further details is available at
www.cambridge.org/PotentialistSetTheory

Appendix A Logico-Structural Potentialism

In this appendix I will fill in the formal details of the Potentialist translation strategy described in Section 6.4.

In Section A.2 I'll define the core kind of structures (standard width initial segments of the hierarchy of sets) which my Potentialist set theory considers the possibility of extending. In Sections A.3 and A.4 and I will show how to cash out claims about it being possible to extend one initial segment V (and choice of some objects x, y, z within that V) to a larger V' containing an object w, without quantifying-in.

A.1 Functional Notation

To avoid overwhelming complexity, we will occasionally resort to functional notation. We now explain how to understand this notation in terms of the language of logical possibility described above.

Definition A.1 R is a function if $(\forall x)(\forall y)(\forall z)(R(x,y) \wedge R(x,z) \rightarrow y = z)$.

So, for example, I would say that admiration is a function (in the actual world) if and only if no one admires two different people. For further notational convenience we will write $f(x_0, \ldots, x_n) = y$ to abbreviate the claim $f(x_0, \ldots, x_n, y)$. When written informally we will understand $f(x_0, \ldots, x_n)$ to stand in for some y such that $f(x_0, \ldots, x_n, y)$.

Finally, given two predicates D and R, I will say f is a function from D to R just if f is a function and $(\forall x)(D(x) \rightarrow (\exists y)f(x,y)) \wedge (\forall x \mid D(x))(\forall y)(f(x,y) \rightarrow R(y))$ and take the notions "surjective," "injective," "domain" and "range" to have their usual meaning.

A.2   Describing Standard-Width Initial Segments

So, let us begin by describing our intended initial segments. Recall the iterative hierarchy conception of sets from Chapter 2. Following Boolos (1971) we imagined a hierarchy of sets consisting of a two sorted structure consisting of:

- a well-ordered series of stages, with no last element; and
- a collection of sets formed at these stages, such that a set is formed at a stage iff its members are all formed at earlier stages.

And we can say that something counts as a standard width initial segment if it satisfies all the requirements above except for the height requirement (that there is no last/highest stage).

I will define a formula $\mathcal{V}(\text{set}, \text{ord}, <, \in, @)$ in the language of logical possibility which expresses the fact that the objects satisfying "ord" are well-ordered by $<$ (giving the well-ordered series of stages), the objects satisfying "set" act like sets under $\in$ and that the relation $@$ relates sets to the stages at which they are formed, so that $@(x, o)$ holds just if the members of x are all available at stages before o. Note that I will sometimes write the relations $<$ and $@$ in infix notation, e.g., $x < y$ rather than $< (x, y)$. Also, I will refer to the elements satisfying "ord" as ordinals and the elements satisfying "set" as sets.

Remember that here that I am using the terms "set," "ord" and "$\in$" for mnemonic and readability purposes alone. As per the Putnamian strategy discussed in Chapter 2, my official Potentialist translation of set theory will employ only logical vocabulary and non-mathematical relations like "is a penciled point," and "there is an arrow from ... to ... ."

Definition A.2 (Initial Segment). The tuple (set, ord, $<, \in, @$) forms an initial segment just if all of the following hold:

1. (ord, $<$) is a well-order[1]
2. $(\forall x)(\forall y)[x \in y \rightarrow \text{set}(x) \wedge \text{set}(y)]$
3. $(\forall x)(\forall y)[@(x, y) \rightarrow \text{set}(x) \wedge \text{ord}(y)]$
4. $(\forall x)(\forall y)[x < y \rightarrow \text{ord}(x) \wedge \text{ord}(y)]$
5. (Fatness) For each o satisfying ord and each way of choosing some elements satisfying set from the sets (i.e., elements satisfying set) available at stages $o' < o$ there is a set with exactly those elements as members:

$$\Box_{\text{set}, \text{ord}, <, \in, @}(\forall o)[\quad \text{ord}(o) \rightarrow$$
$$(\forall x)P(x) \rightarrow \text{set}(x)((\exists o') \wedge \text{ord}(o') \wedge o' < o \wedge (x, o')) \rightarrow$$
$$(\exists y)(\text{set}(y) \wedge @(y, o) \wedge (\forall z)(P(z) \leftrightarrow z \in y))]$$

6. Every set is available at some ordinal level:

---

[1] See Definition E.2 *in Section E of the online appendix for formal definition.*

$$(\forall x)\text{set}(x) \to (\exists o)\text{ord}(o) \land (x, o)$$

7. All sets available at some o such that $\text{ord}(o)$ can only have elements which occur at some level below as elements:

$$(\forall x)(\forall o)(x, o) \to (\forall z)[z \in x(\exists o') \to o' < o \land (z, o')]$$

8. (Extensionality) No two distinct elements satisfying set have exactly the same elements:

$$(\forall x)(\forall y)[\text{set}(x) \land \text{set}(y) \to x = y \lor (\exists z)(\text{set}(z) \land \neg(z \in x \leftrightarrow z \in y)]$$

9. The ordinals are disjoint from the sets:

$$(\forall x)\neg(\text{ord}(x) \land \text{set}(x)).$$

Note that we can think of V_a from the standard set-theoretic hierarchy as corresponding to an initial segment $\langle \text{set}, \text{ord}, <, \in, @ \rangle$ where $\text{ord}, <$ has the same order type as a. Speaking loosely, this means if $@(x, u)$ then x would be in $V_{|u|+1}$ where $|u|$ indicates the ordinal corresponding to u.

I will use $\mathcal{V}(V_i)$ to abbreviate the claim that $\text{set}_i, \in_i$ etc. satisfy the sentence $\mathcal{V}(\text{set}, \text{ord}, <, \in @))$ defined above. I will also use $V(x)$ to abbreviate $\text{ord}(x) \lor \text{set}(x)$ and $\Diamond_V$ abbreviates $\Diamond_{\text{ord}, \text{set}, <, \in, @}$.

A.3 Extensibility

Next, we need to cash out claims about one initial segment extending another.

Definition A.3 (Initial Segment Extension). V_a extends V_b just if all the following hold:

- $\mathcal{V}(V_a)$
- $\mathcal{V}(V_b)$
- $(\forall x)[\text{set}_b(x) \to \text{set}_a(x)]$
- $(\forall x)(\forall y)[\text{set}_b(y) \to (x \in_b y \leftrightarrow x \in_a y)]$
- $(\forall x)[\text{ord}_b(x) \to \text{ord}_a(x)]$
- $(\forall x)(\forall y)[\text{ord}_b(y) \to (x <_b y \leftrightarrow x <_a y)]$
- $(\forall x)(\forall y)[\text{ord}_b(y) \to (x @_b y \leftrightarrow x @_a y)]$

I will use $V_a \geq V_b$ to abbreviate the claim that V_a (i.e., $\text{set}_a, \text{ord}_a, \in_a, @_a, \leq_a$) extends V_b (i.e., $\text{set}_b, \text{ord}_b, \in_b, @_b, <_b$).

If we followed Putnam and Hellman in quantifying-in to the $\Diamond$ of logical possibility, this would suffice to let us write Potentialist translations. We would translate the set theoretic utterance $(\exists x)(\forall y)[\neg x = y \lor \neg y \in x]$ as follows:

$$\Diamond\left((\exists x)\mathcal{V}(V_1) \wedge \text{set}_1(x) \wedge \Box_{V_1}(\forall y)[V_2 \geq V_1 \wedge \text{set}_2(y) \to \neg x = y \vee \neg y \underset{2}{\in} x]\right)$$

In words, it's logically possible there is an initial segment (of the hierarchy of sets) V_1 containing a set x (i.e., $\text{set}_1(x)$) such that its necessary, holding fixed V_1 (i.e., $\text{set}_1, \text{ord}_1, \in_1, <_1, @_1$), that any choice of an element y from a model of set theory V_2 extending V_1 must satisfy $x = y \vee \neg y \in_2 x$.

However, once we embrace the notion of conditional logical possibility, we can banish quantifying-in from our translations and thus avoid certain philosophical controversies.

A.4   Eliminating quantifying-in

The key "trick" which lets us eliminate quantifying-in from our Potentialist paraphrases, will be to supplement out initial segments V_i with a copy of the natural numbers (representing formal variables from the language of set theory) and an assignment function ρ_i which assigns each formal variable (i.e. natural number) to a set (objects satisfying set_i) from V_i. Note that my only reason for using $\mathbb{N}$ is that the natural numbers (under successor) contain infinitely many definable objects, which we can use to represent variables.

Specifically, we represent the natural numbers with the predicate $\mathbb{N}$ and the function S and identify the formal variable x_n with the natural number n (i.e., $\underbrace{S(\ldots S(0))}_{n}$). Rather than use clunky formal variables x_i everywhere we instead use normal letter variables x, y, z, etc. to stand in for particular formal variables and denote the number y stands in for by $\ulcorner y \urcorner$, i.e., if y stands in for x_n then $\ulcorner y \urcorner = n$. We formalize this as follows.

Definition A.4 (Interpreted Initial Segment). Say that the relations in the pair (V, ρ) apply to an interpreted initial segment (written $\vec{\mathcal{V}}(V, \rho)$) just if $\text{set}, \in, \text{ord}, <, @$ satisfy $\mathcal{V}(\text{set}, \in, \text{ord}, <, @)$ and ρ is a function from those objects satisfying $\mathbb{N}$ to those satisfying set. More concretely, this amounts to the conjunction of the following three claims:

- $\mathcal{V}(V)$

- $\mathbb{N}, S$ satisfy $\text{PA}_\Diamond$ (the categorical description of the numbers given in Section J.3 of the online appendix)
- ρ is a function from $\mathbb{N}$ to set.

Note that we prove in Lemma J.11 in Section J.3 of the online appendix that it's logically possible to have $\mathbb{N}, S$ satisfy $\text{PA}_\Diamond$.

I will often use the $\vec{V}_a$ notation to denote the pair V_a, ρ_a. And I will use $\Diamond_{\vec{V}_n}$ to abbreviate claims of the form $\Diamond_{\text{set}_n, \in_n, \text{ord}_n, @_n, \leq_n \rho_n, \mathbb{N}, S}$ (and similarly for $\Box_{\vec{V}_n}$). Note that we use the same relations $\mathbb{N}, S$ for every $\vec{V}_i$.

We can now define a notion of extension for interpreted initial segment.

Definition A.5 (Interpreted Initial Segment Extension). The interpreted initial segment $\vec{V}_b$ extends $\vec{V}_a$ while assigning x written $\vec{V}_a \leq_x \vec{V}_b$ just if:

- $V_a \leq V_b$

- $\mathcal{V}(\vec{V}_a) \wedge \vec{\mathcal{V}}(\vec{V}_b)$

- $(\forall n | \mathbb{N}(n))(\rho_a(n) = \rho_b(n) \vee n = \ulcorner x \urcorner)$

Appendix B   Notation and Some Example Arguments

In this appendix I'll prove some lemmas using only the basic inference rules in Chapter 3.[1] Most importantly, I'll introduce a useful and intuitive way of reasoning about conditional logical possibility: inner $\Diamond$ arguments.

B.1   Inner Diamond

Let us start with the Inner Diamond lemma (Lemma B.1), which will help us capture natural reasoning about conditional logical possibility in a more intuitive manner. Specifically, while the Importing (Axiom 8.6) and Logical Closure (Axiom 8.7) axioms capture the intuition that we can deploy our normal tools of reasoning to infer what further facts must be true in some particular logically possible context, using them directly would force us to carry unwieldy long conjunctions of all facts we've derived are logically possible through our proofs. The Inner Diamond lemma justifies our use of more natural mathematical reasoning.

The intuition behind the Inner Diamond lemma is that reasoning like the following is valid:

Suppose we know the following. There are at least three cats. And it's logically possible, given what cats there are, that every cat is sleeping on a distinct blanket. What else must be true in this logically possible scenario? We can "import" the fact that there are at least three cats (since any scenario which preserves the structural facts about how cathood applies must preserve this fact). So, by first-order logic, this possible scenario must be one in which there are at least three blankets. Thus, it is logically possible, given the facts about what cats there are, that there are at least three blankets.

---

[1] Before beginning, let me note a technical point about how I will talk about lemmas. Consider the trivial lemma whose content is $(\exists x)R(x) \rightarrow (\exists x)R(x)$. We don't regard the proof of this lemma as merely establishing the fact that for some particular relation, e.g., redness, if there is some red thing then there is some red thing. Rather, we regard the lemma as standing in for the fact that this result is provable for *any* one-place relation or, alternately, as proving that the claim in the lemma is logically necessary.

Indeed, we will see shortly if we can prove $(\exists x)R(x) \rightarrow (\exists x)R(x)$ without any premises we can infer $\Box[(\exists x)R(x) \rightarrow (\exists x)R(x)]$ and then (as we will also see below) substitute the relations under the $\Box$ and then eliminate it.

Thus, we allow deducing the fact that $(\exists x)G(x) \rightarrow (\exists x)G(x)$ from the trivial lemma asserting that $(\exists x)R(x) \rightarrow (\exists x)R(x)$. This resembles the situation in first-order logic where we prove that substitution of bound variables preserves truth-value, and then don't pay much attention to the particular bound variables used to express results.

Lemma B.1 (Inner Diamond). *If $\Gamma_1 \vdash \Diamond_{\mathcal{L}} \Theta$ and $\Gamma_2, \Theta \vdash \Phi$, where every element of Γ_2 is a sentence content-restricted to $\mathcal{L}$, then $\Gamma_1, \Gamma_2 \vdash \Diamond_{\mathcal{L}} \Phi$.*

Proof. Consider a scenario where the antecedent of the lemma is true. Assume that Γ_1, Γ_2. Then we have $\Diamond_L \Theta$ by the first assumption. By successive applications Importing (Axiom 8.6) to each of the sentences $\Gamma_2^1, \ldots, \Gamma_2^n$ in Γ_2, we have $\Diamond_L (\Theta \wedge \Gamma_2^1 \wedge \ldots \wedge \Gamma_2^n)$. Now by Logical Closure (Axiom 8.7) and the fact that $\Gamma_2, \Theta \vdash \Phi$ we can get $\Diamond_L \Phi$. Thus $\Gamma_1, \Gamma_2 \vdash \Phi$, as desired. ∎

We note that this lemma supports the following kind of reasoning (as illustrated in the above example).

We derive some sentence of the form $\Diamond_{\mathcal{L}} \Theta$ from the assumptions Γ_1. For instance, in the example above Θ would be the claim that "every cat slept on a distinct blanket" and $\mathcal{L}$ would just be the predicate cat. We then wish to reason about the "world" whose possibility is guaranteed by the fact that $\Diamond_{\mathcal{L}} \Theta$, e.g., the possible world which holds fixed (the structure of) the application of cat and at which every cat slept on a distinct blanket. In that world Θ (every cat slept on a distinct blanket) is true as is, intuitively, every fact content-restricted to cat true in the actual world. For instance, in the example above the fact that there are at least three cats is also true in that world (we refer to the act of taking a sentence content-restricted to $\mathcal{L}$ and concluding it holds at the world whose logical possibility is asserted by $\Diamond_{\mathcal{L}} \Theta$ as importing). We then use proof rules to derive some desired conclusion Φ from Θ and the set of "imported" sentences Γ_2. For instance, in the above example, Φ is the sentence asserting there are at least three blankets. Intuitively, Φ must also be true in the logically possible world under consideration and thus $\Diamond_{\mathcal{L}} \Phi$ must be actually true. In the example above, Γ_2 would just contain the sentence asserting that there are at least three cats. Since $\Theta, \Gamma_2 \vdash \Phi$ and all sentences in Γ_2 are content-restricted to cat this intuition is born out rigorously since the above lemma establishes that $\Gamma_1, \Gamma_2 \vdash \Diamond_{\mathcal{L}} \Phi$.

B.2   Natural Deduction with Inner Diamond Arguments

Since the process of entering $\Diamond_{\mathcal{L}}$ contexts, i.e., using Inner Diamond (Lemma B.1) to reason about what else must be true in a particular logically possible scenario, is unfamiliar and can be a bit tricky, I will informally introduce a natural deduction system for the notion of proof defined in Chapter 8 together with some notational conventions which make it easier to keep track of arguments like the one above (especially in contexts where one must make multiple Inner Diamond arguments within one another).

This system is loosely based around that used by Goldfarb (2003) and I follow his system in citing the line numbers justifying each inference rule to the left of the name of the inference rule, while indicating the assumptions a line depends on by placing those line numbers in brackets (line numbers not in brackets are the lines cited as immediate justification for the current inference). So, for example, we write down Φ 5, 6 X [2, 4, 5] on line 7 of a proof when rule X allows us to conclude Φ from lines

5, 6 and the cumulative set of assumptions from which we've established Φ are the sentences on lines 2, 4 and 5. Note that this system satisfies the principle that if ψ appears on some line of the proof and Γ is the set of sentences appearing on the lines listed in brackets next to ψ then $\Gamma \vdash \psi$.

However, my system differs from Goldfarb's in two primary ways. First, I will allow any purely first-order deduction to be compressed into a single FOL rule. However, I will still sometimes explicitly make use of $\rightarrow I$ to infer $\phi \rightarrow \psi$ in cases where ψ can only be inferred from ϕ via modal reasoning. I will also use Ass. to indicate that a new assumption is being made.

Second, all modal axioms and axiom schema proposed in Chapter 8 are taken to be logical truths. So, any instance of these axiom schemata can be written down with no associated citations or assumptions. And, to save time, any instance of an axiom schema with the form $\phi \rightarrow \psi$ may instead be regarded as an inference rule allowing us to infer ψ from ϕ (citing the line containing ϕ as a justification). For example, this is an acceptable deduction of $\Diamond_P[(\exists x)P(x)] \rightarrow P(x)$:

| 1 | $\Diamond_P[(\exists x)\,P(x)] \rightarrow P(x)$ | $\Diamond$ E |

This is also an acceptable deduction of the same fact:

1	$\Diamond_P[(\exists x)\,P(x)]$	Ass [1]
2	$P(x)$	$1\,\Diamond$ E [1]
3	$\Diamond_P[(\exists x)\,P(x)] \rightarrow P(x)$	$1,2 \rightarrow$ I

Third, and most distinctively, I will introduce a special context called a $\Diamond$ context (nestable to arbitrary depth) corresponding to reasoning via Inner Diamond (Proposition B.1), i.e., reasoning about what else must be true within some scenario which is known to be (conditionally) logically possible. I will graphically indicate what sentences are being asserted or assumed within this context by indentation and a sideways T labeled with a $\Diamond$ to indicate this context.

So, for example, we can represent the following extremely short Inner Diamond argument:

Given what cats and hunters there are, it's logically possible that something is both a cat and a hunter. Any possible situation in which something is both a cat and a hunter, must be a situation in which something is either a cat or a hunter. Therefore, given what cats and hunters there are, its logically possible that something is either a cat or a hunter.

with a proof that looks like this:

1	$\Diamond_{cat}(\exists x)(cat(x) \wedge hunter(x))$	[1]
2 $\Diamond$	$(\exists x)(cat(x) \wedge hunter(x))$ [cat, hunter]	(1) In$\Diamond$ I, [2*]
3	$(\exists x)(cat(x) \vee hunter(x))$	(2) FOL [2*]
4	$\Diamond_{cat}(\exists x)(cat(x) \vee hunter(x))$	(1, 2–3) In$\Diamond$ E [1]

The vertical line going from 2–3 above indicates those lines occur inside a special context. I call this a $\Diamond$ context to indicate that these lines contain reasoning about *what must be true within* a logically possible scenario in which $(\exists x)(cat(x) \wedge hunter(x))$, while all the structural facts about, how cathood applies, are preserved.

What are the rules for writing things down in this context? Recall that the Inner Diamond (Proposition B.1) lemma says that if we have one conditional possibility claim $\Diamond_{\mathcal{L}}(\Theta)$, and some facts Γ_2 which are content-restricted to the relations being held fixed, then if we can show that any such possible scenario where Θ must also be one where Φ (by showing $\Theta, \Gamma_2 \vdash \Phi$), we can infer the corresponding conditional logical possibility clam for Φ.

The key idea will be to use indentation and the Fitch-style sidewise T to graphically distinguish a main line of argument which goes from Γ_2 (where the sentences Γ_2 are content-restricted to $\mathcal{L}$) and $\Diamond_{\mathcal{L}}\Theta$ and then $\Diamond_{\mathcal{L}}\Phi$, from a supporting subproof which shows that $\Gamma_2, \Theta \vdash \Phi$ and thereby justifies the latter inference.

In the latter subproof (which I indent and mark off as a separate context) we are, in essence, attempting to milk new consequences from our knowledge that $\Diamond_{\mathcal{L}}(\Theta)$, by thinking about what *else* must be true in a possible ($\Diamond_{\mathcal{L}}$) situation where Θ. Thus, I will call beginning such a subproof "entering the $\Diamond_{\mathcal{L}}$ context" (associated with some claim $\Diamond_{\mathcal{L}}(\Theta)$ that was established on a previous line), and thereby beginning an Inner Diamond (Lemma B.1) argument.

One will only be permitted to import those claims Γ_2 from the main line of argument (thereby assuming they continue to hold in the current context) which are content-restricted to the relevant list of relation $\mathcal{L}$. And we will only be allowed to close the Inner Diamond context, i.e., dropping back one level of indentation and writing down $\Diamond_{\mathcal{L}}(\Phi)$, if we have proved that Φ holds inside the Inner Diamond context, by showing that it follows from the initial assumption that Θ and some facts Γ_2 which we were allowed to import because they were content-restricted to $\mathcal{L}$.

Reasoning inside a $\Diamond$ context proceeds just as it does normally, with the exception that each line in the context must either be our initial assumption that Θ (where $\Diamond_{\mathcal{L}}\Theta$ is the sentence that opened the diamond context), an instance of "importing" (where the sentence must be imported from the parent context) or be deducible from previous lines within this exact $\Diamond$ context.

While the operation of In $\Diamond$ I is rather straightforward, I'll call attention to one detail. Note that besides line 2 we wrote [2*] rather than [1] as one might expect. We do this to maintain the property that if Ψ is written on a line it is deducible from the lines written in brackets next to it.[2]

---

[2] To this end we treat the initial line in each $\Diamond$ context and every line introduced via importing (see below) as if they were assumptions inside that context. However, we mark these assumptions with an asterisk since they are justified assumptions (it's safe to assume they are true in the $\Diamond$ context) and must be replaced with the line numbers from the parent context when we leave the $\Diamond$ context.

In accordance with this idea, a sentence ρ can be written down inside the "$\Diamond_{\mathcal{L}}$ context" governed by the claim that $\Diamond_{\mathcal{L}}$, iff:

- $\rho = \Theta$
- $\rho = \Psi$ for some Γ which is content-restricted to $\mathcal{L}$ and occurs on an earlier line in the proof which is in the same context as the $\Diamond_{\mathcal{L}}\Theta$ statement used to introduce this Inner Diamond context. (I will, as usual,

One can leave the $\Diamond_{\mathcal{L}}$ context above by going from knowledge that ϕ holds within this context to the conclusion that $\Diamond_{\mathcal{L}}\phi$ holds outside it. We indicate this inference pattern via the rule In $\Diamond$E. This is the only way to introduce a sentence into the current context based on activity in a child context.[3]

B.3   Example of Inner $\Diamond$ with Importing

We can also capture the reasoning in the slightly more complicated argument below, where we use knowledge of suitably content-restricted claims about the actual world to draw consequences from a modal claim.[4]

It's logically possible, given what cats there are, that each slept on a distinct blanket. There are at least three cats. Therefore, it's logically possible, given what cats there are that there are at least three blankets:

1	$\Diamond_{cat}$Each cat slept on a distinct blanket	Ass. [1]
2	There are at least three cats	Ass. [2]
3	$\Diamond$ ⎢ Each cat slept on a distinct blanket	1 In$\Diamond$, [3*]
4	There are more than three cats	2 Import [4*]
5	There are at least three blankets	3,4 FOL [3*,4*]
6	$\Diamond_{cat}$There are at least three blankets	1, 2–5 In$\Diamond$E [1,2]

sometimes elide the steps needed to transform implicitly content-restricted sentences into first-order logically equivalent explicitly content-restricted sentences.)
- ρ follows from previous lines within this $\Diamond$ context by one of the axioms or inference rules for reasoning about logical possibility presented in this book.

[3] Note that In $\Diamond$E may not be applied to any line with uncancelled (unstarred) assumptions introduced in the context being closed. Moreover, In $\Diamond$E must take each starred line number j^* on the line on which Φ appears (here that's line 3 and Φ is $(\exists x)(cat(x) \lor hunter(x))$) and replace it with the assumptions of the line (in the current context) used to justify line j. For instance, in the current case the only (starred) assumption for line 3 is line 2. Looking at line 2 we see that it is justified by reference to line 1 (which is in the current context). So we copy the line numbers in brackets on line 1 into the brackets on line 4 (in this case that's just 1).

[4] Note that the sentence on line 2 "There are at least three cats" is content-restricted to {cats} (assuming that this abbreviates an FOL statement in the usual Fregian fashion). This fact allows us to import it into our reasoning about what the possible scenario where each cat slept on a different blanket must be like on line 4.

Also note that on line 5 we have proved "there are at least three blankets" with only assumptions $[3^*, 4^*]$ (which are starred because they were introduced by Inner Diamond introduction or importing). Thus, we have shown that the conclusion that there are at least three cats follows from things we are entitled to assume about any logically possible scenario witnessing the truth of the sentence on line 1. So we can apply In $\Diamond$ Elimination to complete our Inner Diamond argument, and conclude that $\Diamond_{cat}$ (There are at least three blankets).

Finally, note that the assumption line numbers listed for our conclusion are $[1,2]$. For these are the assumptions needed for the claims about the actual world (namely the sentences on lines 1 and 2), which entitle us to assume that the possible scenario considered on lines 3–5 satisfy the assumptions on lines 3 and 4, which imply there are more than three blankets.

See Appendix G in the online appendix for an explicit formal statement of this natural deduction system, and a demonstration that proofs in it obey the notion of provability above.

B.4   Box Inference Rules

Although the $\Box$ is not an official item in our symbolism, but merely an abbreviation for $\neg\Diamond\neg$, it is often helpful to reason in terms of it. Earlier we proved a couple of rules regarding $\Box$ inferences and here we present several more.

First, I present an introduction rule for $\Box$.

Lemma B.2 ($\Box$ I). *If* $\Gamma \vdash \Theta$ *and every* $\gamma \in \Gamma$ *is a sentence content-restricted to* $\mathcal{L}$ *then* $\Box_{\mathcal{L}}\theta$.

Proof. Suppose for contradiction that Γ, Θ are as above, but the lemma fails, i.e., $\Diamond_{\mathcal{L}}\neg\Theta$. By Inner Diamond (Lemma B.1) with $\Theta_1 = \Diamond_{\mathcal{L}}\neg\Theta$ and $\Theta_2 = \Theta$, we can infer $\Diamond_{\mathcal{L}} \bot$ as $\Theta, \neg\Theta \vdash \Theta \wedge \neg\Theta$. Hence, by $\Diamond$ Elimination (Axiom 8.2) we can export the contradiction. Hence, $\Box_{\mathcal{L}}\Theta$ as desired. ∎ Now I give the corresponding elimination rule.

Lemma B.3 ($\Box$ Elimination). $\Box_{\mathcal{L}}\Theta \rightarrow \Theta$.

Note that I prove a stronger version of this result in Section H of the online appendix that allows arbitrary substitution of relations when eliminating the box and, for ease of reading, I will also refer to that result as $\Box$ Elimination.

Proof. Assume the claim fails. We can derive contradiction immediately by applying $\Diamond$ Introduction (Axiom 8.1) to $\neg\Theta$ to derive $\Diamond_{\mathcal{L}}\neg\Theta$, which is $\neg\Box_{\mathcal{L}}\Theta$. We can write this in terms of the natural deduction system presented above as follows:

1	$\Box_{\mathcal{L}}\Theta$	[Γ]
2	$\neg\Diamond_{\mathcal{L}}\neg\Theta$	[Γ]
3	$\neg\Theta$	Assump. [3]
4	$\Diamond_{\mathcal{L}}\neg\Theta$	4 $\Diamond$ I [3]
5	$\bot$	2, 4 $\bot$ I [3,Γ]
6	$\neg\neg\theta$	3–5 $\neg$I [Γ]
7	θ	6 $\neg$E [Γ] ∎

To give a more visceral sense of how proofs using this logical system work, see Section B.6 where I prove two lemmas which mirror results in set theory (which can be found in elementary texts like Jech (1978)).

B.5 $\Diamond$ Reducing and $\Box$ Expansion

Using first-order logic and the basic principles in Chapter 8 we can prove various useful lemmas.

The Reducing lemma (Lemma B.4) (together with $\rightarrow E$) vindicates intuitive reasoning along the following lines. Suppose it's logically possible, given the facts about friendship and enmity in the actual world, that something has a frenemy (i.e., there are items x and y such that x is the friend of y and x is the enemy of y). Then it's logically possible given (just) the facts about friendship in the actual world that something has a frenemy.

Lemma B.4 (Reducing). *If $\mathcal{L} \supseteq \mathcal{L}'$ then $\Diamond_{\mathcal{L}} \Theta \rightarrow \Diamond_{\mathcal{L}'} \Theta$,*

Proof. First note that if $\mathcal{L} \supseteq \mathcal{L}'$ then any sentence of the form $\Diamond_{\mathcal{L}'} \Theta$ is content-restricted to $\mathcal{L}$.

Assume that $\Diamond_{\mathcal{L}} \Theta$. We have $\Theta \vdash \Diamond_{\mathcal{L}'} \Theta$, by $\Diamond$ Introduction (Axiom 3.1). So, by Logical Closure (Axiom 3.7), we have $\Diamond_{\mathcal{L}}(\Diamond_{\mathcal{L}'} \Theta)$. Then by $\Diamond$ Elimination (Axiom 3.2) we can conclude that $\Diamond_{\mathcal{L}'} \Theta$ (since $\Diamond_{\mathcal{L}'} \Theta$ is content-restricted to $\mathcal{L}$). Thus, we have $\Diamond_{\mathcal{L}} \Theta \rightarrow \Diamond_{\mathcal{L}'} \Theta$. ∎

We note that this immediately entails a corresponding expansion property for sentences under the $\Box$.

Lemma B.5 (Box Expanding). *If $\mathcal{L}' \supset \mathcal{L}$ then $\Box_{\mathcal{L}} \Phi \rightarrow \Box_{\mathcal{L}'} \Phi$.*

Proof. Assume that $\Box_{\mathcal{L}} \Phi$ and suppose for contradiction that $\neg \Box_{\mathcal{L}'} \Phi$, hence $\Diamond_{\mathcal{L}'} \neg \Phi$. By the Reducing lemma (Lemma B.4) we can infer $\Diamond_{\mathcal{L}} \neg \Phi$, which contradicts our assumption that $\Box_{\mathcal{L}} \Phi$. ∎

B.6   Lemmas about Well-Orderings

To give a more visceral sense of how proofs using my logical system work, I'll now prove two lemmas which mirror results in set theory (which can be found in elementary texts like Jech (1978)). In each case, I will make an argument verbally, and then follow it up with an argument using the formal notation (making explicit when we enter and leave Inner Diamond contexts).

Elsewhere I will present proofs in a more informal style. However, I hope the completely explicit proofs in this section will help the reader understand how these informal proofs can be expanded into a formal argument.

B.6.1   Reconstructing Well-Orderings: Part I

Jech's version of the first lemma I am going to prove says the following:

"If $(W, <)$ is a well-ordered set and $f : W \rightarrow W$ is an increasing function, then $x < f(x)$ for each $\in W$." (Jech 1978)

We can write a version of Jech's Lemma follows (see Section E of the online appendix for the definition of a well-order):

Lemma B.6 *If f is an embedding of the well-order $W, <$ into itself then* $(\forall x, y : W(x) \wedge W(y))(x < f(x))$.

We define embedding as follows:

Definition B.1 (Definition of Embedding). A two-place relation f **is an embedding** of $W, <$ into $W', <'$ iff:

- f is a function (remember we define what it takes for a relation to qualify as a function in Section A.1)
- $(\forall x)[W(x) \rightarrow (\exists y)(W'(y) \wedge f(x))]$ i.e., f maps all of W into W'
- $(\forall x)(\forall y)(\forall x')(\forall y')[x < y \leftrightarrow f(x) < f(y)]$, i.e., f respects $<$.

Remember that we've defined function so that the function $f(x)$ is a convenient way of talking about the relation $f(x, y)$ satisfying $f(x, y) \wedge f(x', y) \rightarrow x = x'$.

As usual, I will sometimes abbreviate the claim that $x < y \vee x = y$ as $x \leq y$.

Proof. To prove this, we will use essentially the same reasoning which Jech uses to prove his set theoretic version of this claim.

Assume that f is an embedding of $(W, <)$ into itself, as per the statement of the lemma. And suppose, for contradiction, the lemma fails. As in Jech's proof, our aim will be to use the properties of well-orderings to derive the existence of a $<$ least counterexample, i.e., an x in W such that $\neg x < f(x) \wedge (\forall y : y < x)(y < f(y))$ and derive contradiction from this.

Applying Simple Comprehension (Axiom 8.4) to $\neg x < f(x)$ tells us it would logical possible – while holding fixed the facts about how $W, <, f$ apply in the situation we are currently considering – for the predicate G apply to just such counterexamples. That is:

$$\Diamond_{W, <, f}(\forall x)(G(x) \leftrightarrow \neg x < f(x))$$

Now we can enter this $\Diamond_{W, <, f}$ context, i.e., begin an Inner Diamond (Lemma B.1) argument, where we reason about what *else* must be true in a possible scenario where (the facts about $W, <, f$ in our original scenario are held fixed but) we also have:

$$(\forall x)(G(x) \leftrightarrow \neg x < f(x))$$

Now the premises of the lemma (that f is an embedding of $(W, <)$ into itself and $(W, <)$ a well-order) and the assumption that the conclusion of the lemma fails are all implicitly context restricted to $W, <, f$ (seen by appropriately restricting all the quantifiers). So, all of these statements must all remain true in this new context and can by imported into this context.

Thus, we can infer that G is non-empty, from the assumption that the lemma fails, i.e., $(\exists x \mid W(x))(\neg x < f(x))$, together with the fact that $(\forall x)(G(x) \leftrightarrow \neg x < f(x))$.

We know that $W, <$ is a well-ordering, and the least element condition from the definition of well-ordering (Definition E.2 in the online appendix) says the following:

$$\Box_{W, <}[(\exists x \mid W(x))G(x) \rightarrow (\exists y \mid W(y) \wedge G(y))(\forall z \mid W(z) \wedge G(z))(y \leq z)]$$

So, by $\square$ Elimination (Lemma B.3) we can infer the existence of a least counterexample y, i.e.,

$$(\exists y \mid W(y) \wedge G(y))(\forall z \mid W(z) \wedge G(z)))(y \leq z))$$

Now let $z = f(y) < y$. By our assumption that f is an embedding (and thus must respect $<$) it follows from $z < y$ that $f(z) < f(y) = z$. So, by the equation specifying the extension of G above we can infer that $G(z)$. Thus, z is an satisfies $G(z)$ and is less than y. Contradiction $\perp$.

Exiting the above $\lozenge$ context (i.e., completing our Inner Diamond argument), we get $\lozenge_{W,<,f}\perp$. And from this $\perp$ follows by $\lozenge$ Elimination (Axiom 8.2) (remembering that $\perp$ is content-restricted to the empty list). Hence, the desired conclusion follows by contradiction. ∎

Intuitively speaking, the argument above shows that the if there were a counterexample to the lemma then it would be logically possible (indeed logically possible, while holding fixed the $W, <, f$ facts!) for the canonical contradiction $\perp$ to be true. But it's not logically possible for $\perp$ to be true. So, there is no counterexample to the lemma.

Representing this proof in terms of our natural deduction system:

1	f is an embedding of the well-order $(W, <)$ into itself	[1]
2	$(\exists x \mid W(x))(f(x) < x)$	[2]
3	$\lozenge_{W,<,f}\forall z[G(z) \leftrightarrow f(z) < z]$	3 Simple Comprehension
4 $\lozenge$	$\forall z[G(z) \leftrightarrow f(z) < z]$ $\{W, <, f\}$	3 In$\lozenge$ I [4*]
5	f is an embedding of the well-order $W, <$ into itself	1 imp [5*]
6	$(\exists x \mid W(x))(f(x) < x)$	2 imp [6*]
7	$\square_{W,<}[(\exists x \mid W(x) \wedge G(x)) \rightarrow (\exists y)[W(y) \wedge G(y) \wedge (\forall z)(W(z) \wedge G(z) \rightarrow y \leq z)])]$	5 FOL [5*]
8	$(\exists x \mid W(x) \wedge G(x)) \rightarrow (\exists y)[W(y) \wedge G(y) \wedge (\forall z)(W(z) \wedge G(z) \rightarrow y \leq z)])$	7 $\square$E [5*]
9	$\perp$	4, 6, 8 FOL [4*,5*,6*]
10	$\lozenge_{W,<,f}\perp$	3, 4–9 In$\lozenge$E [1,2]
11	$\perp$	10 $\lozenge$E [1,2]
12	$\neg(\exists x)(f(x) < x)$	11 (2) FOL [1]

B.6.2   Reconstructing Well-Orderings: Part 2

Jech (1978) writes, "No well-ordered set is isomorphic to an initial segment of itself." We can state the claim to be proved using the definition of isomorphism (Definition 7.4) from Chapter 7.

Lemma B.7 *If $(W, <)$ is a well-ordering and there is some x in W such that W' applies to just those $z < x$ in W then $\neg\lozenge_{W,W',<}\langle W, >\rangle \underset{f}{\cong} \langle W', >\rangle$.*

Proof. Let $W, W', <$ be as in the lemma and suppose for contradiction that $\Diamond_{W,W',<}\langle W, >\rangle \cong_f \langle W', >\rangle$. Using Inner Diamond (Lemma B.1) we can enter this $\Diamond$ context. We can import the fact $W, <$ is a well-order (because it is content-restricted to W, W' and $<$). By first-order logic and unpacking definitions we can infer from the fact that f isomorphically maps $\langle W, >\rangle$ to $\langle W', >\rangle$ that f is an embedding of $W, <$ into $W', <$. And, by the assumptions about W' above, this implies that f is an embedding of $W, <$ into $, <$.

Now, to get contradiction, note that by Lemma B.6 (all instances of which are provable from empty premises, hence provable in any $\Diamond$ context) f does not map any object satisfying W strictly $<$-below itself. However, we know there is an object x satisfying W which is $>$ all objects satisfying W' and that $\langle W, >\rangle \cong_f \langle W', >\rangle$. It follows by first-order logic that f maps this x to some $y < x$. Thus, we have derived contradiction/the false ($\bot$) from premises which would have to obtain in this (supposedly) logically possible scenario.

As before, we can conclude this inner $\Diamond_{W,W',<}$ argument and returning to our original context with the conclusion that $\Diamond_{W,W',<}\bot$. And from this $\bot$ follows by $\Diamond$ Elimination (Axiom 8.2).

This completes our proof by contradiction that there can be no f isomorphically mapping $\langle W, >\rangle$ to a proper initial segment of itself. ∎

We can use the natural deduction system to expose the modal reasoning within this argument, as follows:

1   $(W, <)$ is a well-ordering and W' is a proper initial segment of W   Ass. [1]

2   $\Diamond_{W,W',<}\langle W, >\rangle \cong_f \langle W', >\rangle$ Ass. [2]

3   $\Diamond$ | $\langle W, <\rangle \cong_f \langle W', <\rangle$ $\{W, W', >\}$ 2 In$\Diamond$I [3*]

4 | $(W, <)$ is a well-ordering and W' is a proper initial segment of W   1 import [4*]

5 | f is an embedding of the well-order $(W, <)$ into itself   3,4 FOL [3*, 4*]

6 | $\neg(\exists x)(f(x) < x)$ 6 Lemma A

7 | $\bot$ 3, 4, 6 FOL [3*,4*]

8   $\Diamond_{W,W',>}(\bot)$ 2, 3–7 In$\Diamond$E [1,2]

9   $\bot$ 8 $\Diamond$E [1,2]

10   $\neg\Diamond_{W,W',<}\langle W, >\rangle \cong_f \langle W', >\rangle$ 9 (2) FOL [1]

B.7   Pasting and Collapsing

Finally, I will conclude this chapter with two lemmas involving more complex modal reasoning using multiple $\Diamond$ contests. The first lemma tells us when two logically possible facts can be inferred to be jointly possible.

One cannot generally infer from $\Diamond_{\mathcal{L}}\Phi$ and $\Diamond_{\mathcal{L}}$ to $\Diamond_{\mathcal{L}}(\Phi \wedge \Psi)$. Consider the case where Φ says there are exactly eight million things and Ψ says there are exactly nine million things. However, the Pasting lemma (Lemma B.8) says that one *can* make this inference in the special situation when the sentences Φ and Ψ are content-restricted so that they can only make claims about the objects satisfying some disjoint lists of relations $\mathcal{I}$ and $\mathcal{J}$ (and how these relate the actual $\mathcal{L}$-structure, which both $\Diamond_{\mathcal{L}}\Phi$ and $\Diamond_{\mathcal{L}}\Psi$ preserve).

Lemma B.8 (Pasting). *Let* $\mathcal{I}$, $\mathcal{J}$ *and* $\mathcal{L}$ *be pairwise disjoint sets of relations. If* $\Diamond_{\mathcal{L}}\Phi$, *where* Φ *is content-restricted to* $\mathcal{L}, \mathcal{I}$ *and* $\Diamond_{\mathcal{L}}\Psi$, *where* Ψ *is content-restricted to* $\mathcal{L}, \mathcal{I}$, *then* $\Diamond_{\mathcal{L}}(\Phi \wedge \Psi)$.

Intuitively speaking, the facts about content-restriction above ensure that attempting to make the sentences inside both possibility claims true at the same time cannot impose conflicting demands. For the only relations whose extensions are relevant to the truth of both sentences are the relations on the list $\mathcal{L}$. And our assumptions say that it's possible to make each interior sentence true while fixing the actual application of these relations.

Proof. Let Φ be content-restricted to $\mathcal{L}, \mathcal{I}$ and Ψ to $\mathcal{L}, \mathcal{J}$, as per the antecedent:

1	$\Diamond_{\mathcal{L}}\Phi$		Ass. [1]
2	$\Diamond_{\mathcal{L}}\Psi$		Ass. [2]
3	$\Diamond$ Φ $\{\mathcal{L}\}$		1, In$\Diamond$I [3*]
4	$\Diamond_{\mathcal{L}}\Psi$		2 import [4*]
5	$\Diamond_{\mathcal{L},\mathcal{I}}\Psi$		4 $\Diamond$ Ignoring [4*]
6	$\Diamond$ Ψ $\{\mathcal{L},\mathcal{I}\}$		5 In$\Diamond$I [6*]
7	Φ		3 import [7*]
8	$\Phi \wedge \Psi$		6,7 FOL [6*, 7*]
9	$\Diamond_{\mathcal{L},\mathcal{I}}(\Phi \wedge \Psi)$		5, 6–8 In$\Diamond$E [3*,4*]
10	$\Diamond_{\mathcal{L}}(\Phi \wedge \Psi)$		9 Reducing [3*, 4*]
11	$\Diamond_{\mathcal{L}}(\Diamond_{\mathcal{L}}(\Phi \wedge \Psi))$		1, 3–10 In$\Diamond$E [1,2]
12	$\Diamond_{\mathcal{L}}(\Phi \wedge \Psi)$		11 $\Diamond$E [1,2]

Informally, this deduction corresponds to the following reasoning. Assume that $\Diamond_{\mathcal{L}}\Phi$ and $\Diamond_{\mathcal{L}}\Psi$. We can prove our claim by making two nested Inner Diamond (Lemma B.1) arguments.

First enter the $(\Diamond_{\mathcal{L}})$ context associated with $\Diamond_{\mathcal{L}}\Phi$. That is, consider what else must be true in any such possible $(\Diamond_{\mathcal{L}})$ situation where Φ. In this situation $\Diamond_{\mathcal{L}}\Psi$ must remain true, for it is content-restricted to $\mathcal{L}$, and we are considering a scenario which preserves the $\mathcal{L}$ facts. By $\Diamond$ Ignoring (Axiom 8.3) it follows that $\Diamond_{\mathcal{L},\mathcal{I}}\Psi$.

Now enter this second, interior, $\Diamond_{\mathcal{L},\mathcal{I}}$ context. That is, consider what must be true in a further possible scenario where Ψ is true while all facts about how relations $\mathcal{L},\mathcal{I}$ applied in the scenario we previously considered are preserved. Here we clearly have Ψ. But we can import the fact that Φ from the previous context, because it is content-restricted to $\mathcal{L},\mathcal{I}$. So we can deduce $\Phi \wedge \Psi$.

Now, leaving this inner $\Diamond_{\mathcal{L},\mathcal{I}}$ context, we can conclude that $\Diamond_{\mathcal{L},\mathcal{I}}(\Phi \wedge \Psi)$. And we can infer that that $\Diamond_{\mathcal{L}}(\Phi \wedge \Psi)$ by $\Diamond$ Ignoring (Axiom 8.3) (because $\mathcal{L}$ is clearly a sublist of $\mathcal{L},\mathcal{I}$).

So, leaving the larger $\Diamond_{\mathcal{L}}$ context we can conclude that $\Diamond_{\mathcal{L}}(\Diamond_{\mathcal{L}}(\Phi \wedge \Psi))$ holds in the situation we were originally considering.

Finally, because $\Diamond_{\mathcal{L}}(\Phi \wedge \Psi)$ is content-restricted to $\mathcal{L}$, we can use $\Diamond$E to draw the desired conclusion $\Diamond_{\mathcal{L}}(\Phi \wedge \Psi)$. ∎

The other lemma concerns when we can collapse multiple logical possibility operators into a single operator.

Lemma B.9 (Diamond Collapsing). *If* $\mathcal{L}' \supseteq \mathcal{L}$ *then* $\Diamond_{\mathcal{L}}\Diamond_{\mathcal{L}'}\Phi \leftrightarrow \Diamond_{\mathcal{L}}\Phi$.

Proof. To prove the left to right direction, suppose that $\Diamond_{\mathcal{L}}\Diamond_{\mathcal{L}'}\Phi$. Enter the $\Diamond_{\mathcal{L}}$ context. In this context we have $\Diamond_{\mathcal{L}'}\Phi$. Since $\mathcal{L}' \supseteq \mathcal{L}$, by Reducing (Lemma B.4) we can infer $\Diamond_{\mathcal{L}}\Phi$. Exiting the $\Diamond_{\mathcal{L}}$ context, we have $\Diamond_{\mathcal{L}}(\Diamond_{\mathcal{L}}\Phi)$ in our original contest. So, we can apply $\Diamond$ Elimination (Axiom 8.2) to infer $\Diamond_{\mathcal{L}}\Phi$.

1	$\Diamond_{\mathcal{L}}\Diamond_{\mathcal{L}'}\Phi$	[1]
2	$\Diamond$ | $\Diamond_{\mathcal{L}'}\Phi$	1 In $\Diamond$ I [2*]
3	| $\Diamond_{\mathcal{L}}\Phi$	2 Reducing [2*]
4	$\Diamond_{\mathcal{L}}(\Diamond_{\mathcal{L}}\Phi)$	1,2–3 In $\Diamond$ E [1]
5	$\Diamond_{\mathcal{L}}\Phi$	4 $\Diamond$ Elimination [1]
6	$\Diamond_{\mathcal{L}}\Diamond_{\mathcal{L}'}\Phi \to \Diamond_{\mathcal{L}}\Phi$	5 → I

To prove the other direction, suppose that $\Diamond_{\mathcal{L}}\Phi$. Entering this diamond context, we have Φ and can infer that $\Diamond_{\mathcal{L}'}\Phi$ by $\Diamond$ Introduction (Axiom 8.1). So, completing our Inner Diamond argument gives us $\Diamond_{\mathcal{L}}\Diamond_{\mathcal{L}'}\Phi$.

1	$\Diamond_{\mathcal{L}}\Phi$	[1]
2	$\Diamond$ | Φ	[$\mathcal{L}$] 1 In $\Diamond$ I [2*]
3	| $\Diamond_{\mathcal{L}'}\Phi$	3, $\Diamond$ I [2*]
4	$\Diamond_{\mathcal{L}}(\Diamond_{\mathcal{L}'}\Phi)$	1, 2–3 In $\Diamond$ E [1]

We also observe that there is a corresponding □ version of the above lemma.

Lemma B.10 (Box Collapsing). *If $\mathcal{L}' \supseteq \mathcal{L}$ then $\Box_{\mathcal{L}} \Phi \leftrightarrow \Box_{\mathcal{L}} \Box_{\mathcal{L}'} \Phi$.*

Proof. Note that this is equivalent to proving

$$\neg \Box_{\mathcal{L}} \Phi \leftrightarrow \neg \Box_{\mathcal{L}} \Box_{\mathcal{L}'} \Phi$$

which is just

$$\Diamond_{\mathcal{L}} \neg \Phi \leftrightarrow \Diamond_{\mathcal{L}} \Diamond_{\mathcal{L}'} \neg \Phi$$

This is true by Diamond Collapsing (Lemma B.9). ∎

Appendix C   Vindication of FOL Inference in Set Theory

In Chapter 9 I show that Potentialist translations of all the ZFC axioms are true and can be justified within my formal system. However, this is not enough to justify ordinary mathematical practice. We also need to show that everything set theorists *derive* from the ZFC axioms using FOL has a true and justified Potentialist translation. And this fact is not immediately guaranteed by the soundness of first-order logic, because our Potentialist translations of set theoretic sentences have a different logical form from the originals.

In this appendix I will show that, for any two sentences of Actualist set theory, if ϕ first-order logically implies that ψ then $t(\phi)$ implies $t(\psi)$. Specifically, I will show that, for every first-order logical proof in the language of set theory, there is a corresponding proof from my inference rules for logical possibility which takes us from the translation of the premises for this argument to the translation of its conclusion. That is, I will prove Theorem 9.1 from Chapter 9, which I restate below.

Theorem 9.1 (Logical Closure of Translation). *Suppose* Φ, Ψ *are sentences in the language of set theory and* $\Phi \vdash_{FOL} \Psi$ *then* $t(\Phi) \vdash t(\Psi)$

Or, equivalently, if $\vdash_{FOL} \Phi \rightarrow \Psi$ then $\vdash t(\Phi) \rightarrow t(\Psi)$.

C.1   Proof Strategy

I will prove the above result by first establishing a more general result about Potentialist translations of arbitrary set theoretic formulae (below), which implies the fact we want about sentences. Note that any formula we are translating should be assumed to be in the language of set theory.

Proposition C.1. *Given a set* Γ *of formulas in the language of set theory if* $\Gamma \vdash_{FOL} \theta$ *then* $\vec{\mathcal{V}}(V_n), t_n(\Gamma) \vdash t_n(\theta)$, *where* $t_n(\Gamma)$ *denotes the pointwise image of* Γ *under* t_n.

Remember that $t_n(\Phi)$ intuitively represents the translation of Φ with respect to the structure V_n and the assignment function ρ_n. Thus, this theorem can be thought of as showing that first-order inferences are valid even with respect to the partial translation t_n.

As one might expect, this more general result implies the Logical Closure of Translation, as we prove below.

Proof. Consider any Φ, Ψ such that $\vdash_{FOL} \Phi \to \Psi$. It follows that $\Phi \vdash_{FOL} \Psi$ and by the theorem above, we know that $\vec{\mathcal{V}}(\vec{V}_n), t_n(\Phi) \vdash t_n(\Psi)$ and thus $\vdash \vec{\mathcal{V}}(\vec{V}_n) \to (t_n(\Phi) \to t_n(\Psi))$.

Now assume that $t(\Phi)$. By this is just $(\vec{\mathcal{V}}(\vec{V}_0) \to t_0(\Phi))$. From this we may infer $\vec{\mathcal{V}}(\vec{V}_0) \to t_0(\Phi)$ and by using the fact that $\vec{\mathcal{V}}(\vec{V}_0) \to (t_0(\Phi) \to t_0(\Psi))$ we can conclude $\vec{\mathcal{V}}(\vec{V}_0) \to t_0(\Psi)$. So we have $t(\phi) \vdash \vec{\mathcal{V}}(\vec{V}_0) \to t_0(\Psi)$.

Since $\vec{\mathcal{V}}(\vec{V}_0) \to (t_0(\Phi) \to t_0(\Psi))$ is provable from empty premises we also have $t(\Phi) \vdash \vec{\mathcal{V}}(\vec{V}_0) \to t_0(\Psi)$. So by $\square$ Introduction (Lemma B.2) and the fact that $t(\Phi)$ is content-restricted to the empty sentence, we can infer $t(\Phi) \vdash (\vec{\mathcal{V}}(\vec{V}_0) \to t_0(\Psi))$. Hence $t(\Phi) \vdash t(\Psi)$ and thus $\vdash t(\Phi) \to t(\Psi)$. $\blacksquare$

We now prove Proposition C.1 via structural induction on first-order proofs (note that technically this is a meta-theorem and the induction occurs in our meta-language). However, first we need a formal definition of an FOL proof.

The choice to explicitly define a notion of proof (as opposed to simply defining the set of provable sentences) might seem odd here. After all, it would be mathematically more elegant to simply define provability as the smallest relation closed under certain rules. However, defining an explicit notion of proof allows us to induct on proof length in establishing the above proposition rather than trying to define some kind of well-founded relation on sequents.

We think of proofs in terms of the familiar tree structure, but formalize this notion in a way which makes it clear what rule is being applied at each point, as below.

Definition C.1 (First-order Proof). $\Gamma \vdash_{FOL} \theta$ just if there is a first-order proof of θ from $\Gamma = \gamma_1, \ldots \gamma_m$ where this is inductively defined as follows (taking the various rule names are understood to refer to distinct constants[1]) and $\langle \ldots \rangle$ to denote an ordered tuple.)

If $\theta \in \Gamma$ then $\langle Ass, \theta \rangle$ is a proof of θ from Γ.

If $\theta = \phi \wedge \psi$ and P_ϕ is a proof of ϕ from Γ and P_ψ is a proof of ψ from Γ then $\langle \wedge I, P_\phi, P_\psi, \phi \wedge \psi \rangle$ is a proof of θ from Γ.

If $\theta = \phi$ or $\theta = \psi$ and $P_{\phi \wedge \psi}$ is a proof of $\phi \wedge \psi$ from Γ then $\langle \wedge E, P_{\phi \wedge \psi}, \theta \rangle$ is a proof of θ from Γ

If $\theta = \phi \vee \psi$ and P is a proof of ϕ from Γ or ψ from Γ then $\langle \vee I, P, \phi \vee \psi \rangle$ is a proof of θ from Γ.

If $P_{\phi \vee \psi}$ is a proof of $\phi \vee \psi$ from Γ and $P_{\phi \vdash \theta}$ is a proof of θ from Γ, ϕ and $P_{\psi \vdash \theta}$ is a proof of θ from Γ, ψ, then $\langle \vee E, P_{\phi \vee \psi}, P_{\phi \vdash \theta}, P_{\psi \vdash \theta}, \theta \rangle$ is a proof of θ from Γ.

If $\theta = \phi \to \psi$ and P_ψ is a proof of ψ from $\Gamma \cup \phi$ then $\langle \to I, P_\psi, \phi \to \psi \rangle$ is a proof of θ from Γ.

If $P_{\phi \to \theta}$ is a proof of $\phi \to \theta$ from Γ and P_ϕ is a proof of ϕ from Γ then $\langle \to E, P_{\phi \to \theta}, P_\phi, \theta \rangle$ is a proof of θ from Γ.

If $\theta = \neg \phi$ and P_ψ is a proof of ψ from Γ, ϕ and $P_{\neg \psi}$ is a proof of $\neg \psi$ from Γ, ψ then $\langle \neg I, P_\psi, P_{\neg \psi}, \neg \phi \rangle$ is a proof of $\neg \phi$ from Γ.

---

[1] For instance, numbers if formalized in an arithmetic meta-language.

If $P_{\neg\neg\theta}$ is a proof of $\neg\neg\theta$ from Γ then $\langle DNE, P_{\neg\neg\theta}\theta\rangle$ is a proof of θ from Γ.

If $\theta = (\forall v)\phi$ and P_ϕ is a proof of ϕ from some $\Gamma' \subseteq \Gamma$ with v not free in any member of Γ' then $\langle \forall I, P_\phi, (\forall v)\phi\rangle$ is a proof of θ from Γ.

If $\theta = \phi(v|v')$ where v' is free for v in θ^2 and $P_{(\forall v)\phi}$ is a proof of $(\forall v)\phi$ from some Γ then $\langle \forall E, P_{(\forall v)\phi}, \theta\rangle$ is a proof of θ from Γ.

If $\theta = v = v$, where v is any variable then $\langle (= I), \theta\rangle$ is a proof of θ from Γ.

If θ is obtained from ϕ by replacing zero or more occurrences of v_1 with v_2, provided that no bound variables are replaced, and all substituted occurrences of v_2 are free and $P_=$ is a proof of $v_1 = v_2$ from Γ and P_ϕ is a proof of ϕ from Γ then $\langle (= E), P_=, P_\phi, \theta\rangle$ is a proof of θ from Γ.

If $P_{\psi\wedge\neg\psi}$ is a proof of $\psi\wedge\neg\psi$ from Γ then $\langle \perp I, P_{\psi\wedge\neg\psi}, \perp\rangle$ is a proof of $\perp$ from Γ.

($\perp$ E) If $\theta = \neg\phi$ and $P_\perp$ is a proof of $\perp$ from Γ then $\langle \perp I, P_\perp, \theta\rangle$ is a proof of θ from Γ.

Note that there is no conflict between our definition of $\forall x$ as an abbreviation of $\neg\exists x\neg$ and our use of the introduction and elimination rules for $\forall$ rather than $\exists$ in proofs (the rule $\forall E$ simply applies to statements of the form $\neg\exists v\neg\psi$).

Definition C.2 If P is a first-order proof then P' is a subproof of P just if either:

- P has the form $\langle R, P_0, \theta\rangle$ and $P' = P_0$ or P' is a subproof of P_0
- P has the form $\langle R, P_0, P_1, \theta\rangle$ and P' is P_0 or P_1 or a subproof of P_0 or P_1.

C.2   Proof of Main Result

We are now in a position to prove Proposition C.1.

Proof. Suppose that $\vec{V}_n$ is an interpreted initial segment, $t_n(\gamma)$ holds for all $\gamma \in \Gamma$ and P is a proof of θ from Γ. Furthermore, assume, by way of induction, that the proposition holds for all P' a subproof of P. We prove that $t_n(\theta)$ also holds (which by the inductive hypothesis demonstrates that $\vec{V}(\vec{V}_n), t_n(\Gamma) \vdash t_n(\theta)$).

Now consider the possible cases for P:

$$P = \langle Ass, \theta\rangle$$

In this case we have $\theta \in \Gamma$ so we immediately have $\vec{V}(V_n), t_n[\Gamma] \vdash t_n(\theta)$:

$P = \langle R, \ldots\rangle$ where $R \in \wedge I, \wedge E, \vee I, \vee E, \rightarrow I, \rightarrow E, \neg I, DNE$

This follows immediately from the fact that t_n commutes with truth-functional operations and the validity of the above rules in our system for reasoning about logical possibility. For example, if $\langle \wedge I, P_\phi, P_\psi, \phi\wedge\psi\rangle$ where $\theta = \phi\wedge\psi$ then $t_n(\phi\wedge\psi)$ would be $t_n(\phi)\wedge t_n(\psi)$ and by the inductive assumption applied to P_ϕ, P_ψ we know that $t_n(\phi)$ and $t_n(\psi)$ both hold, yielding the desired conclusion:

---

[2] That is, if substituting v with v' does not lead to any variable which was antecedently free becoming bound. Here $\theta(v|v')$ stands for the result of substituting all free instances of v in θ with instances of v'.

$$P = \langle (=I), \theta \rangle$$

In this case $t_n(\theta)$ is $\rho_n(\ulcorner v \urcorner) = \rho_n(\ulcorner v \urcorner)$, which trivially follows from the assumption that V_n is an interpreted initial segment (hence ρ_n is functional with $\ulcorner v \urcorner$ in its domain):

$$P = \langle (=E), P_=, P_\phi, \theta \rangle$$

By applying the inductive hypothesis to $P_=$ we have $\vdash t_n(v_1 = v_2)$, which is $\rho_n(v_1) = \rho_n(v_2)$. By the inductive hypothesis applied to P_ϕ we can infer $t_n(\phi)$. As θ is obtained from ϕ by replacing zero or more occurrences of v_1 with some v_2 (where no bound variables are replaced and all substituted occurrences of v_2 are free) the Variable Swap lemma (Lemma L.5 in Section L.2 of the online appendix) lets us deduce $t_n(\theta)$.

The proof of the Variable Swap lemma (Lemma L.5) can be found in Section L.2 of the online appendix but it should be intuitively clear that if $\rho_n(v_1) = \rho_n(v_2)$ then replacing some number of occurrences of v_1 in θ with v_2 can't change its truth-value:

$$P = \langle \forall I, P_\phi, \theta \rangle$$

By our definition of First-order Proof (Definition C.1) $\theta = (\forall v)\phi(v)$ for some formula ϕ and variable v, and P_ϕ is a proof of ϕ from some $\Gamma' \subset \Gamma$ containing no formula γ in which v appears free.

If $\gamma \in \Gamma'$ then by the inductive hypothesis applied to Γ', P_ϕ and $n+1$ we have $\vec{\mathcal{V}}(\vec{V}_n), t_n[\Gamma] \vdash t_n(\gamma)$. Now suppose $\vec{V}_{n+1} \geq_v \vec{V}_n$. As v is not free in γ, by the Translation theorem we can prove that $t_{n+1}(\gamma)$. Thus $\vec{\mathcal{V}}(\vec{V}_n), t_n[\Gamma] \vdash \vec{V}_{n+1} \geq_v \vec{V}_n \rightarrow t_{n+1}(\phi)$.

The Translation theorem (Theorem L.1) is proved[3] in Section L.2 of the online appendix. However, it intuitively says that the truth-value of $t_n(\gamma)$ only depends on how $\vec{V}_n$ assigns the free variables in γ and not on the height of $\vec{V}_n$. Thus, if $\vec{V}_m \geq_v \vec{V}_n$ and v isn't free in γ then we can infer $t_m(\gamma)$ from $t_n(\gamma)$.

As, by Lemma 7.1, every member $t_n[\Gamma]$ is content-restricted to $\vec{V}_n$. Thus, by $\Box$ Introduction (Lemma B.2) we may deduce that

$$t_n((\forall v)\phi(v)) \overset{\text{def}}{\leftrightarrow} \Box_{V_n} \vec{V}_{n+1} \underset{v}{\geq} \vec{V}_n \rightarrow t_{n+1}(\phi)$$

$$P = \langle \forall E, P_{\forall v\phi}, \theta \rangle$$

By definition of First-order Proof (Definition C.1), θ is equal to $\phi(v|v')$ for some formula ϕ and variable v where none of the substituted instances of v' are bound and $P_{\forall v\phi}$ is a proof of $\forall v\phi$.

To prove this claim we merely need to show that if every logically possible way of extending $\vec{V}_n$ with $\vec{V}_{n+1}$ and choosing v makes $t_{n+1}(\phi)$ true then whatever assignment $\vec{V}_n$ makes for v' makes $t_n(\phi(v|v'))$ true. To this end we must use the intuitive fact that

---

[3] Note that Hellman proves something analogous to this lemma in Hellman (1996), assuming there are infinitely many inaccessibles (but I make no such assumption).

whatever assignment $\vec{V}_n$ makes for v' there is some extension $\vec{V}_{n+1}$ which makes the same assignment for v. This fact is proved in the Pointwise Interpretation Tweaking lemma (Lemma L.1 in Section L of the online appendix).

We first note that by Pointwise Interpretation Tweaking lemma we have

$$\Diamond_{\vec{V}_n}\left(\vec{V}_{n+1}\underset{v}{\geq}\vec{V}_n\wedge\rho_{n+1}(\ulcorner v\urcorner)=\rho_n(\ulcorner v'\urcorner)\right)$$

Enter this $\Diamond_{\vec{V}_n}$ context. By Lemma 7.1 each sentence in $t_n[\Gamma]$ can be inferred to remain true in this context.[4] So, by the inductive hypothesis applied to $P_{\forall v\phi}$ we may infer

$$t_n((\forall v)\phi(v))\overset{\text{def}}{\leftrightarrow}\Box_{V_n}(\vec{V}_{n+1}\underset{v}{\geq}\vec{V}_n\rightarrow t_{n+1}(\phi))$$

Application of $\Box$ Elimination (Lemma 7.4) allows us to infer $t_{n+1}(\phi)$ and from there, as $\rho_{n+1}(\ulcorner v\urcorner)=\rho_n(v')=\rho_{n+1}(v')$ we may apply the Variable Swap lemma (lemma Lemma L.5 in Section L.2 of the online appendix) to derive $t_{n+1}(\theta)$.

As $\theta=\phi(v|v')$ if v' isn't v then v doesn't appear free in θ. If v' is v then $\rho_{n+1}(v')=\rho_n(v')$ and in either case as $\vec{V}_{n+1}\geq_v\vec{V}_n$ we have that ρ_{n+1} and ρ_n agree on all free variables in θ. Hence by the Translation theorem (Theorem L.1 in the online appendix) we can infer $t_n(\theta)$. Leaving the $\Diamond_{\vec{V}_n}$ context we have $\Diamond_{\vec{V}_n}t_n(\theta)$. Since by Lemma 2.1 $t_n(\theta)$ is content-restricted to $\vec{V}_n$ by $\Diamond$ Elimination (Axiom 8.2) we can conclude $t_n(\theta)$.∎

C.3   Justifying Truth Condition Adequacy

I claim that if our Platonist paraphrase satisfies the Definable Supervenience Condition and captures intended truth conditions for all sentences in S then the if-thenist paraphrase strategy above does as well. That is, we can produce a nominalistic sentence which is true at all the same possible worlds where the Platonist would say their logically regimented sentence is true – exactly the possible worlds where we think the English sentence being regimented is intuitively true:

Paraphrase: $\Box_N(D\rightarrow\Phi_P)$ is true at a world w iff the Platonist would say Φ_p. For suppose that the above conditions are satisfied for some applied mathematical sentence Φ and N, and D and P. I claim that Platonist must admit that if Φ is true iff $\Box_N(D\rightarrow\Phi)$ For, consider any metaphysically possible world w. The Platonist thinks D is true at w, by the fact that they take D to be metaphysically necessary. But then, by the following theorem we have Φ iff my translation of Φ (i.e., $\Box_N(D\rightarrow\Phi)$) is true.

---

[4] Strictly speaking we only need to import the finitely many $\gamma\in\Gamma$ used to prove $\forall v\phi$.

C.3.1   Formal Justification for Truth Conditions Adequacy

Theorem C.1 *Suppose that:*

1. $\mathcal{P} \cap \mathcal{N} = \varnothing$
2. *both Φ_p and D are content-restricted[5] to $\mathcal{P} \cup \mathcal{N}$*
3. *D is a categorical description of $\mathcal{P}$ over $\mathcal{N}$,*

$$\text{then} \left[D \rightarrow \left(\Phi \leftrightarrow T(\phi) \right) \right]$$

Remember that $T(\phi) = \square_{\mathcal{N}}(D \rightarrow \phi)$.

Note that while all theorems proved in this book hold with necessity, we make the necessity claim explicit here as it is used to justify claim that $T(\phi)$ matches the truth-value the Platonist intended ϕ to have at every metaphysically possible world.

Proof. Note that it is enough to prove $D \rightarrow (\Phi \leftrightarrow T(\phi))$ as we may then invoke $\square$ Introduction (Lemma B.2) to infer the necessity of the claim. So we suppose that D holds and verify $\Phi \leftrightarrow T(\phi)$.

($\leftarrow$) Suppose $T(\phi) = \square_{\mathcal{N}}(D \rightarrow \Phi)$. By $\square$ Elimination (Lemma B.3) we may infer $D \rightarrow \Phi$ and thus Φ.

($\rightarrow$) Suppose, for a contradiction, that Φ holds but $T(\phi)$ fails to hold. Thus, we can infer, $\Diamond_{\mathcal{N}}(D \wedge \neg \Phi)$. Letting $\mathcal{P}'$ be a set of new relations of the same arity as $\mathcal{P}$ and applying Relabeling (Axiom 8.5) we may infer $\square_{\mathcal{N}}(D[\mathcal{P}/\mathcal{P}'] \wedge \neg \Phi[\mathcal{P}/\mathcal{P}'])$ (where $\mathcal{P}/\mathcal{P}'$ indicates simultaneously replacing the relations in $\mathcal{P}$ by those in $\mathcal{P}'$).

Now since we are assuming that both D and Φ hold we may infer $\Diamond_{\mathcal{N}}(D \wedge \Phi)$ via $\Diamond$ Introduction (Axiom 8.1). As D and Φ are both content-restricted to $\mathcal{N} \cup \mathcal{P}$ (and thus $D[P/P] \wedge \neg \Phi[P/P]$ is content-restricted to $\mathcal{N} \cup \mathcal{P}$) we may use Pasting (Lemma B.7) to infer

$$\Diamond_{\mathcal{N}} D \wedge D[\mathcal{P}/\mathcal{P}'] \wedge \Phi \wedge \neg \Phi[\mathcal{P}/\mathcal{P}']$$

Enter this $\Diamond_{\mathcal{N}}$ context and import the following fact (content-restricted to the empty set) from the definition of categorical over (Definition 12.3):

$$(D[N_1, \ldots, N_m, P_1, , P_n] \wedge D[N_1, \ldots, N_m, P_1/P'_1, \ldots, P_n/P'_n] \rightarrow \mathcal{N} \cup \mathcal{P} \cong \mathcal{N} \cup \mathcal{P})$$

By application of (full) $\square$ Elimination (Lemma H.4 of Section H of the online appendix) we can infer $D \wedge D[\mathcal{P}/\mathcal{P}'] \rightarrow \mathcal{N} \cup \mathcal{P} \cong \mathcal{N} \cup \mathcal{P}'$ and thus $\mathcal{N} \cup \mathcal{P} \cong \mathcal{N} \cup \mathcal{P}'$. Hence, we can use the Isomorphism theorem (Theorem I.1 of Section I of the online appendix) to infer from this and the fact that Φ holds that $\Phi[\mathcal{P}/\mathcal{P}']$ holds. But this contradicts the fact that $\neg \Phi[\mathcal{P}/\mathcal{P}']$. Exporting this contradiction gives us the desired conclusion. ∎

---

[5] Intuitively, given any D that is a categorical description of $\mathcal{P}$ over $\mathcal{N}$ it should be possible to find a D' that imposes the same restrictions but is content-restricted to $\mathcal{P} \cup N$. However, it is not clear how to go about proving this.

Appendix D Archimedean and Rich Instantiation

It is easy to show that the platonic structures appealed to here (the natural numbers and the functions from the natural numbers to paths) definably supervene on the how nominalistic relations apply, via the techniques used in Section 12.2):

Platonist "x has length a finite multiple of y": for all n there are paths c_i, $1 \leq i \leq n$ with $c_0 = x$, $c_1 = y$ and $c_i \oplus x = c_{i+1}$.

Then we can prove the uniqueness Putnam claims under the following nominalistically stateable assumptions, which may[1] imply that space is infinite in extent:

- given a path x there are paths y with length equal to any finite multiple[2] of x;
- no path is infinite in length with respect to another, i.e., if $x \leq_L y$ then some finite multiple of x is longer[3] than y;
- the relations $\leq_L$, $\oplus_L$ have the basic properties you would expect from their role as length comparisons.[4]

That is, the assumptions above imply that there is a unique (up to multiplicative constant) length function (from paths to the real numbers) respecting $\leq_L$, $\oplus_L$. Hence there's a unique length ratio function $l_r(p_1, p_2) = r$ such that for all functions f satisfying the above constraints, $f(p_1)/f(p_2) = r$.

One can also (as seems more in the spirit of Putnam's cryptic remark about how his uniqueness claim is to be proved) establish the same uniqueness on the assumption that space is (roughly) infinitely divisible rather than infinite in extent, replacing the finite multiple condition by a finite division condition, i.e., for each path y and finite multiple

---

[1] It's not entirely clear that the Closure Under Multiples requirement requires that space is infinite in extent. For instance, if space is (as some models of General Relativity would suggest) closed back on itself (i.e., has the geometry of the surface of a four-dimensional sphere) then sufficiently long paths would simply start wrapping around the universe. Assuming one is willing to accept such paths then it is much more plausible that something like these conditions hold necessarily (not an assumption we must vindicate but see below) as one might think space can't have an edge.

[2] The fact that y_2 is twice the length of x can be expressed as $\oplus_L(x, x, y)$, the fact that y_3 is three times the length of x can be expressed as the conjunction of the claim that y_2 is twice the length of x and $\oplus_L(x, y_2, y_3)$. The closure condition simply asserts the existence of y_i for each y (note that while this is a schema, we can replace this with a single formula expressing the same condition in the language of logical possibility by using a categorical description of the natural numbers and turning this into a statement about this logically possible structure).

[3] Formally, any time the length of a path a is less than the length of a path b there are paths c_i, $1 \leq i \leq n$ with length i times that of a and c_n is longer than b. Again, logical possibility allows us to formalize this schema with an equivalent single sentence.

[4] For instance $\leq_L$ is transitive, reflexive, etc. and $\oplus_L(p_1, p_2, p_3) \leftrightarrow \oplus_L(p_2, p_1, p_3)$, etc.

m there is a path *x* such that *m* copies of *x* have the same length as *y*. And we can write a corresponding division rather than multiplication-based version of the Archimedean principle. Philosophers who have a more classical view of space might find such a condition more plausible as a necessary constraint on the nature of space (we will also consider the possibility that neither seems necessary below as well):

- given a path *x* there are paths *y* with length equal to any finite multiple of *x*;
- no path is infinite in length with respect to another, i.e., if $x \leq_L y$ then *x* is longer than some finite divisor[5] of *y*;
- the relations $\leq_L$, $\oplus_L$ have the basic properties you would expect from their role as length comparisons.[6]

Standard measurement theoretic uniqueness arguments then show that if either of these two claims are satisfied then "length is **richly instantiated**" in the following sense:

$\Box_{\leq_L, \oplus_L, path}[D \rightarrow$ If *f* and *g* are functions satisfying Putnam's measurement theoretic axioms for being a length function, then *f* and *g* agree up to a constant][7]

So, we know that if length is richly instantiated in a world *w* then we have uniqueness and the Platonist paraphrase above yields the correct truth conditions at that world.

D.1   Inferential Role Adequacy

We can also show that the nominalistic paraphrase strategy produced by our translation *T* preserves the desired inferential role of scientific sentences in *S*, capturing both inferences between scientific sentences and inferences between scientific sentences and observational sentences (on the plausible assumption that the latter can be understand as content restricted to some nominalist vocabulary).

It is easy to see that applying our translation *T* preserves inference relations between scientific statements in the following sense.

Theorem D.1 *Suppose that Φ, Ψ are content restricted to $\mathcal{P} \cup \mathcal{N}$ and $\vdash \Phi \rightarrow \Psi$ then $\vdash T(\Phi) \rightarrow T(\Psi)$. Furthermore, if $\vdash T(\Phi) \rightarrow T(\Psi)$ then $\vdash (D \wedge \Phi) \rightarrow \Psi$.*

Proof. As $\vdash \Phi \rightarrow \Psi$, we have $(D \rightarrow \Phi) \vdash (D \rightarrow \Psi)$ by FOL. Suppose $\Box_{\mathcal{N}}(D \rightarrow \Phi)$, then by the above fact and Box Closure (Lemma H.2 of Section H of the online appendix) we have $\Box_{\mathcal{N}}(D \rightarrow \Psi)$.

---

[5] Formally, anytime the length of a path *a* is less than the length of a path *b* there are paths c_i, $1 \leq i \leq n$ with length *i* times that of *a* and c_n is longer than *b*. Again, logical possibility allows us to formalize this schema with an equivalent single sentence.

[6] For instance $\leq_L$ is transitive, reflexive, etc. and $\oplus_L(p_1, p_2, p_3) \leftrightarrow \oplus_L(p_2, p_1, p_3)$, etc.

[7] Recall that *D* says there are (objects with the structure of) numbers and functions from spatial paths to numbers and some functions.

Furthermore, suppose that $D \wedge \Phi$. Above we proved that $D \rightarrow (\Phi \leftrightarrow T(\Phi))$ holds for all sentences content restricted to $\mathcal{P} \cup \mathcal{N}$. Thus, we can infer $T(\Phi)$ and $T(\Psi)$, so Ψ. Thus, $(D \wedge \Phi) \rightarrow \Psi$. ∎

Note the "furthermore" ensures that this translation strategy doesn't let you prove any more than the Platonist thinks the scientist can prove.

We can also show the following theorem:

Theorem D.2 *Suppose that the conditions for T being defined above are satisfied (specifically $\Diamond_\mathcal{N} D$, i.e., the Platonist isn't supposing the existence of incoherent objects), and that Φ is content restricted to* N *then* $\Phi \leftrightarrow T(\Phi)$.

Proof. ($\rightarrow$) Assume Φ. Hence, we may infer $D \leftrightarrow \Phi$. As this proof only assumed Φ which is content restricted to N we may infer $T(\Phi) = \Box_\mathcal{N}[D \leftrightarrow \Phi]$ via $\Box$ Introduction (Lemma B.2).

($\leftarrow$) Assume $T(\Phi) = \Box_\mathcal{N}[D \leftrightarrow \Phi]$. By assumption we have $\Diamond_\mathcal{N} D$. Enter this $\Diamond_\mathcal{N}$ context. As $T(\Phi)$ is content restricted to N we may import it. By $\Box$ Elimination (Lemma B.3) we may infer $D \leftrightarrow \Phi$ and hence Φ.

Leaving this Inner Diamond context gives us $\Diamond_\mathcal{N} \Phi$. As Φ content restricted to $\mathcal{N}$ we may conclude Φ by $\Diamond$ Elimination (Axiom 8.2). ∎

This fact shows that our translation $T(\phi)$ of a platonistic theory ϕ implies the same nominalistic sentences observation sentences as the original theory does when combined with the Platonist's assumption that D (on the plausible assumption that all observation sentences are nominalistic, i.e., content restricted to nominalistic vocabulary). For in this case, we have $T(\phi) \leftrightarrow \psi$ iff $T(\phi) \rightarrow T(\phi)$ by Theorem D.2 iff $(D \wedge \phi) \leftrightarrow \psi$ by Theorem D.1. So, our translation of a platonistic scientific theory ϕ implies a nominalistic consequence iff ϕ itself implies ψ given the platonist's assumptions D about relevant mathematical objects existing.

Bibliography

Azzouni, Jody. 2003. *Deflating Existential Consequence*. Oxford University Press.

Baker, Alan. 2005. "Are There Genuine Mathematical Explanations of Physical Phenomena?" *Mind* 114 (454): 223–38.

2016. "Parsimony and Inference to the Best Mathematical Explanation." *Synthese* 193 (2): 333–50. https://doi.org/10.1007/s11229-015-0723-3

Benacerraf, Paul. 1965. "What Numbers Could Not Be." *The Philosophical Review* 74 (1): 47–73. https://doi.org/10.2307/2183530

1973. "Mathematical Truth." *Journal of Philosophy* 70: 661–80.

Benacerraf, Paul, and Hilary Putnam. 1983. *Philosophy of Mathematics Selected Readings*. Cambridge University Press.

Berry, Sharon. 2015. "Chalmers, Quantifier Variance and Mathematicians' Freedom." In *Quantifiers, Quantifiers, and Quantifiers. Themes in Logic, Metaphysics and Language*, edited by Alessandro Torza, 191–219. Springer.

2018a. "Modal Structuralism Simplified." *Canadian Journal of Philosophy* 48 (2): 200–22. https://doi.org/10.1080/00455091.2017.1344502

2018b. "(Probably) Not Companions in Guilt." *Philosophical Studies* 175 (9): 2285–308.

2019a. "Gunk Mountains: A Puzzle." *Analysis* 79 (1): 3–10.

2019b. "Quantifier Variance, Mathematicians' Freedom and the Revenge of Quinean Indispensability Worries." *Erkenntnis* September. https://doi.org/10.1007/s10670-020-00298-1

2020. "Coincidence Avoidance and Formulating the Access Problem." *Canadian Journal of Philosophy*.

2021. "Physical Possibility and Determinate Number Theory." *Philosophia Mathematica* nkab013. https://doi.org/10.1093/philmat/nkab013

Bliss, Ricki, and Kelly Trogdon. 2016. "Metaphysical Grounding." In *The Stanford Encyclopedia of Philosophy*, edited by Edward N. Zalta, Winter. Metaphysics Research Lab, Stanford University. https://plato.stanford.edu/archives/win2016/entries/grounding/

Boghossian, Paul Artin. 1996. "Analyticity Reconsidered." *Noûs* 30 (3): 360–91. https://doi.org/10.2307/2216275

Boolos, George. 1971. "The Iterative Conception of Set." *Journal of Philosophy* 68 (8): 215–31.

1985. "Nominalist Platonism." *Philosophical Review* 94 (3): 327–44. https://doi.org/10.2307/2185003

1989. "Iteration Again." *Philosophical Topics* 17 (2): 5–21. https://doi.org/10.5840/philtopics19891721

Bueno, Otávio. 2020. "Nominalism in the Philosophy of Mathematics." In *The Stanford Encyclopedia of Philosophy*, edited by Edward N. Zalta, Fall. Metaphysics Research

Lab, Stanford University. https://plato.stanford.edu/archives/fall2020/entries/nominal ism-mathematics/

Burali-Forti, C. 1897. "Una Questione Sui Numeri Transfiniti." *Rendiconti Del Circolo Matematico Di Palermo (1884–1940)* 11 (1): 154–64. https://doi.org/10.1007 /BF03015911

Burgess, John P. 2018. "Putnam on Foundations: Models, Modals, Muddles." In *Hilary Putnam on Logic and Mathematics*, 129–43. Springer International Publishing. https://books .google.com/books?id=sxwkuAEACAAJ

Burgess, John P., and Rosen, Gideon 1997. *A Subject with No Object: Strategies for Nominalistic Interpretation of Mathematics*. Oxford University Press.

Button, Tim. n.d. *Open Set Theory.* www.homepages.ucl.ac.uk/~uctytbu/OERs.html

Button, Tim, and Sean Walsh. 2016. "Structure and Categoricity: Determinacy of Reference and Truth Value in the Philosophy of Mathematics." *Philosophia Mathematica* 24 (3): 283–307.

Chalmers, David J. 2009. "Ontological Anti-Realism." In *Metametaphysics: New Essays on the Foundations of Ontology*, edited by David John Chalmers, David Manley, and Ryan Wasserman. Oxford University Press.

2012. *Constructing the World*. Oxford University Press.

Chrisley, R., and S. Begeer. 2000. *Artificial Intelligence: Critical Concepts*. Vol. 1. Taylor & Francis.

Clarke-Doane, Justin. 2012. "Moral Epistemology: The Mathematics Analogy." *Nous*.

2020. *Morality and Mathematics*. Oxford University Press.

Cohen, Paul J. 1963. "The Independence of the Continuum Hypothesis." *Proceedings of the National Academy of Sciences of the United States of America* 50 (6): 1143–48.

Cole, Julien. 2013. "Towards an Institutional Account of the Objectivity, Necessity, and Atemporality of Mathematics." *Philosophia Mathematica* 21(1): 9–36.

Colyvan, Mark. 2001. *The Indispensability of Mathematics*. Oxford University Press.

2010. "There Is No Easy Road to Nominalism." *Mind* 119 (474): 285–306. https://doi.org/10 .1093/mind/fzq014

2019. "Indispensability Arguments in the Philosophy of Mathematics." In *The Stanford Encyclopedia of Philosophy*, edited by Edward N. Zalta, Spring. Metaphysics Research Lab, Stanford University. https://plato.stanford.edu/archives/spr2019/entries/mathphil-indis/

Davidson, Donald. 1967. "Truth and Meaning." *Synthese* 17 (1): 304–23.

Dummett, Michael. 1991. *Frege: Philosophy of Language*, Vol. 1: 995. Harvard University Press.

1993. *The Seas of Language*, Vol. 58. Oxford University Press.

Eddon, M. 2013. "Quantitative Properties." *Philosophy Compass* 8 (7): 633–45.

Einheuser, Iris. 2006. "Counterconventional Conditionals." *Philosophical Studies* 127: 459–82.

Eklund, Matti. 2009. "Carnap and Ontological Pluralism." In *Metametaphysics: New Essays on the Foundations of Ontology*, edited by David J. Chalmers, David Manley, and Ryan Wasserman, 130–56. Oxford University Press.

Etchemendy, John. 1990. *The Concept of Logical Consequence*. Harvard University Press.

Feferman, Solomon. 2011. "Is the Continuum Hypothesis a Definite Mathematical Problem?" (Draft) http://logic.harvard.edu/EFI_Feferman_IsCHdefinite.pdf [Accessed 20 September 2021].

Field, Hartry. 1980. *Science Without Numbers: A Defense of Nominalism*. Princeton University Press.

1984. "Is Mathematical Knowledge Just Logical Knowledge?" *Philosophical Review*, 93 (4): 509–52.

1989. *Realism, Mathematics & Modality*. Blackwell.

1998. "Which Undecidable Mathematical Sentences Have Determinate Truth Values?" *Truth in Mathematics*, 291–310.

2008. *Saving Truth from Paradox*. Oxford University Press.

Fine, Kit. 1984. "Critical Review of Parsons' Non-Existent Objects." *Philosophical Studies* 45 (1): 95–142.

2005. "Our Knowledge of Mathematical Objects." In *Oxford Studies in Epistemology*, edited by T. Z. Gendler and J. Hawthorne, 89–109. Clarendon Press.

2006. "Relatively Unrestricted Quantification." In *Absolute Generality*, edited by Agustín Rayo and Gabriel Uzquiano, 20–44. Oxford University Press.

Frege, Gottlob. 1980. *The Foundations of Arithmetic: A Logico-Mathematical Enquiry into the Concept of Number*. Northwestern University Press.

Gaifman, Haim. 2012. "On Ontology and Realism in Mathematics." *Review of Symbolic Logic* 5 (3): 480–512. https://doi.org/10.1017/s1755020311000372

Garson, James. 2016. "Modal Logic." In *The Stanford Encyclopedia of Philosophy*, edited by Edward N. Zalta, Spring. Metaphysics Research Lab, Stanford University. https://plato.stanford.edu/archives/spr2016/entries/logic-modal/

Gödel, Kurt. 1931. "Über Formal Unentscheidbare Sätze Der Principia Mathematica Und Verwandter Systeme, I." *Monatshefte für Mathematik Und Physik* 38: 173–98.

1947. "What Is Cantor's Continuum Problem?" In *Kurt Gödel: Collected Works,* Vol. II, 176–187. Oxford University Press.

Goldfarb, Warren. 2003. *Deductive Logic*. Hackett Publishing Company.

Goles, Eric, Oliver Schulz, and Mario Markus. 2001. "Prime Number Selection of Cycles in a Predator-Prey Model." *Complexity* 6 (4): 33–8. https://onlinelibrary.wiley.com/doi/abs/10.1002/cplx.1040

Goodman, Nelson. 1955. *Fact, Fiction, and Forecast*. Harvard University Press.

Gómez-Torrente, Mario. 2000. "A Note on Formality and Logical Consequence." *Journal of Philosophical Logic* 29 (5): 529–39. https://doi.org/10.1023/A:1026510905204

Hamkins, Joel. 2012. "The Set-Theoretic Multiverse." *The Review of Symbolic Logic* 5 (3): 416–49. https://doi.org/10.1017/S1755020311000359

2013. "The Multiverse Perspective in Set Theory." Harvard University. http://jdh.hamkins.org/wp-content/uploads/2013/08/Harvard-2013-Summary.pdf

Hanson, William H. 2006. "Actuality, Necessity, and Logical Truth." *Philosophical Studies* 130 (3): 437–59. https://doi.org/10.1007/s11098-004-5750-8

Hellman, Geoffrey. 1994. *Mathematics Without Numbers*. Oxford University Press.

1996. "Structuralism Without Structures." *Philosophia Mathematica* 4 (2): 100–23.

1998. "Maoist Mathematics?", *Philosophia Mathematica*, 6 (3): 334–45, https://doi.org/10.1093/philmat/6.3.334

2011. "On the Significance of the Burali-Forti Paradox." *Analysis* 71 (4): 631–37. https://doi.org/10.1093/analys/anr091

2020. private communication.

Hilbert, David. 1926. "Über Das Unendliche." *Mathematische Annalen* 95 (1): 161–90. https://doi.org/10.1007/BF01206605

Hirsch, Eli. 2010. *Quantifier Variance and Realism: Essays in Metaontology*. Oxford University Press.

Horsten, Leon. 2019. "Philosophy of Mathematics." In *The Stanford Encyclopedia of Philosophy*, edited by Edward N. Zalta, Spring. Metaphysics Research Lab, Stanford University. https://plato.stanford.edu/archives/spr2019/entries/philosophy-mathematics/

Jech, Thomas. 1978. *Set Theory*. Academic Press.

   1981. "Set Theory." *Journal of Symbolic Logic*, 876–77.

Jenkins, Carrie. 2008. *Grounding Concepts: An Empirical Basis for Arithmetical Knowledge*. Oxford University Press.

Jonas, Silvia. 2020. "Mathematical and Moral Disagreement." *Philosophical Quarterly* 70 (279): 302–27. https://doi.org/10.1093/pq/pqz057

Koellner, Peter. 2009. "On Reflection Principles." *Annals of Pure and Applied Logic* 157 (2–3): 206–19.

   2016. "Infinity Up on Trial: Reply to Feferman." *Journal of Philosophy* 113 (5/6): 247–60. https://doi.org/10.5840/jphil20161135/616

   n.d. "Hamkins on the Multiverse." https://pdfs.semanticscholar.org/26fb/06dc48cfe92d32dd bd8fd963047687b4459d.pdf

Kripke, Saul A. 1963. "Semantical Considerations on Modal Logic." *Acta Philosophica Fennica* 16 (1963): 83–94.

Lewis, D. K. 1990. "Noneism or Allism?" *Mind* 99 (393): 23–31.

   1991. *Parts of Classes*. Basil Blackwell.

Linnebo, Øystein. 2010. "Pluralities and Sets." *Journal of Philosophy* 107 (3): 144–64.

   2013. "The Potential Hierarchy of Sets." *Review of Symbolic Logic* 6 (2): 205–28.

   2018a. *Thin Objects*. Oxford University Press.

   2018b. "Putnam on Mathematics as Modal Logic." In *Hilary Putnam on Logic and Mathematics*, 249–167. Springer International Publishing.

Lyon, Aidan. 2012. "Mathematical Explanations of Empirical Facts, and Mathematical Realism." *Australasian Journal of Philosophy* 90 (3): 559–78. https://doi.org/10.1080/00048402.2011.596216

Martin, Donald. n.d. "Completeness or Incompleteness of Basic Mathematical Concepts." http://math.ucla.edu/~dam/booketc/efi.pdf.

   2001. "Multiple Universes of Sets and Indeterminate Truth Values." *Topoi* 20 (1): 5–16. https://doi.org/10.1023/A:1010600724850

McGee, Vann. 1997. "How We Learn Mathematical Language." *Philosophical Review* 106 (1): 35–68.

Melia, Joseph. 1995. "On What There's Not." *Analysis* 55 (4): 223–29.

Nolt, John. 2018. "Free Logic." In *The Stanford Encyclopedia of Philosophy*, edited by Edward N. Zalta, Fall. Metaphysics Research Lab, Stanford University. https://plato.stanford.edu/archives/fall2018/entries/logic-free/

Parsons, Charles. 1977. "What Is the Iterative Conception of Set?" In *Logic, Foundations of Mathematics, and Computability Theory: Part One of the Proceedings of the Fifth International Congress of Logic, Methodology and Philosophy of Science, London, Ontario, Canada-1975*, edited by Robert E. Butts and Jaakko Hintikka, 335–67. Springer. https://doi.org/10.1007/978-94-010-1138-9_18

   2005. *Mathematics in Philosophy*. Cornell University Press.

   2007. *Mathematical Thought and Its Objects*. Cambridge University Press.

Potter, Michael. 2004. *Set Theory and Its Philosophy: A Critical Introduction*. Oxford University Press.

2007. "What Is the Problem of Mathematical Knowledge?" In *Mathematical Knowledge*, edited by Michael Potter, Mary Leng, and Alexander Paseau.

Prior, Arthur. 1960. "The Runabout Inference Ticket." *Analysis* 21 (December): 38–39.

Pruss, Alexander R. 2019. "Might All Infinities Be the Same Size?" *Australasian Journal of Philosophy* 98 (3): 604–17. https://doi.org/10.1080/00048402.2019.1638949

Putnam, Hilary. 1967. "Mathematics Without Foundations." *Journal of Philosophy* 64 (1): 5–22.

1971. *Philosophy of Logic*. Allen & Unwin.

1983. "Models and Reality." *In Realism and Reason*. Cambridge University Press.

2000. "Paradox Revisited II: Sets—A Case of All or None." In *Between Logic and Intuition: Essays in Honor of Charles Parsons*, edited by Gila Sher and Richard L. Tieszen, 16–26. Cambridge University Press.

Quine, W. V. O. 1953. "Reference and Modality." In *Journal of Symbolic Logic*, edited by Willard O. Quine, 137–38. Harvard University Press.

1960. "Carnap and Logical Truth." *Synthese* 12 (4): 350–74.

1961. "On What There Is." In *From a Logical Point of View*, edited by W. V. Quine, 1–19. Cambridge, Mass.: Harvard University Press.

1970. *Philosophy of Logic*. Harvard University Press.

Rayo, Agustín. 2015. *The Construction of Logical Space*. Oxford University Press.

Reinhardt, William. 1974. "Remarks on Reflection Principles, Large Cardinals, and Elementary Embeddings." *Proceedings of Symposia in Pure Mathematics* 10: 469–205.

Resnik, Michael D. 1981. "Mathematics as a Science of Patterns: Ontology and Reference." *Noûs* 15 (4): 529–50. https://doi.org/10.2307/2214851

Rizza, David E. 2011. "Magicicada, Mathematical Explanation and Mathematical Realism." *Erkenntnis* 74 (1): 101–14. https://doi.org/10.1007/s10670-010-9261-z

Roberts, Sam. 2017. "A Strong Reflection Principle." *Review of Symbolic Logic* 10 (4): 651–62.

2018. "Modal Structuralism and Reflection." *The Review of Symbolic Logic*, 1–38.

Rumfitt, Ian. 2015. *The Boundary Stones of Thought: An Essay in the Philosophy of Logic*. Oxford University Press.

Scanlon, T. M. 2014. *Being Realistic About Reasons*. Oxford University Press.

Schwarz, Wolfgang 2013. Contingent Identity. *Philosophy Compass* 8 (5):486–95.

Searle, John. 1995. *The Construction of Social Reality*. Free Press.

Shapiro, Stewart, and Teresa Kouri Kissel. 2018. "Classical Logic." In *The Stanford Encyclopedia of Philosophy*, edited by Edward N. Zalta, Spring. Metaphysics Research Lab, Stanford University. https://plato.stanford.edu/archives/spr2018/entries/logic-classical/

Shapiro, Stewart, and Crispin Wright. 2006. "All Things Indefinitely Extensible." In *Absolute Generality*, edited by Agustín Rayo and Gabriel Uzquiano. 255–304. Clarendon Press.

Shapiro, Stuart. 1997. *Philosophy of Mathematics: Structure and Ontology*. Oxford University Press,.

Sider, Theodore. 2009. "Ontological Anti-Realism." In *Metametaphysics: New Essays on the Foundations of Ontology*. Oxford University Press.

2011. *Writing the Book of the World*. Oxford University Press.

n.d. "A Crash Course on Measurement Theory." https://tedsider.org/teaching/structuralism_18/crash%20course%20on%20measurement%20theory.pdf

Studd, J. P. 2019. *Everything, More or Less: A Defence of Generality Relativism*. Oxford University Press.

Tait, William. 2005. *The Provenance of Pure Reason*. Oxford University Press.

Tennant, Neil. 2017. "Logicism and Neologicism." In *The Stanford Encyclopedia of Philosophy*, edited by Edward N. Zalta, Winter. Metaphysics Research Lab, Stanford University. htt ps://plato.stanford.edu/archives/win2017/entries/logicism/

Thomasson, Amie L. 2015. *Ontology Made Easy*. Oxford University Press .

Uzquiano, Gabriel. 1996. "The Price of Universality." *Philosophical Studies*.

Wang, Hao. 1998. "A Logical Journey. From Gödel to Philosophy." *Philosophy* 73 (285): 495–504.

Warren, Jared. 2014. "Quantifier Variance and the Collapse Argument." *Philosophical Quarterly* 65 (259): 241–53.

Williamson, Timothy. 2008. *The Philosophy of Philosophy (The Blackwell/Brown Lectures in Philosophy)*. Wiley-Blackwell.

  2013. *Modal Logic as Metaphysics*. Oxford University Press.

Index

Printed in the United States
by Baker & Taylor Publisher Services